Property Tables, Figures, and Charts
to Accompany

THERMODYNAMICS
AN ENGINEERING APPROACH
Second Edition

YUNUS A. CENGEL
Department of Mechanical Engineering
University of Nevada, Reno

MICHAEL A. BOLES
Department of Mechanical and Aerospace Engineering
North Carolina State University

McGraw-Hill, Inc.

New York   St. Louis   San Francisco   Auckland   Bogotá
Caracas   Lisbon   London   Madrid   Mexico City   Milan   Montreal
New Delhi   San Juan   Singapore   Sydney   Tokyo   Toronto

This book is printed on recycled, acid-free paper containing a minimum of 50% recycled de-inked fiber.

Property Tables, Figures, and Charts to Accompany
THERMODYNAMICS: An Engineering Approach
Second Edition

ISBN 0-07-011222-3

34567890 BKM BKM 909876

Property Tables, Figures, and Charts (SI Units)

Same as one shown

$\frac{kPa \cdot m^3}{kg \cdot K}$

TABLE A-1
Molar mass, gas constant, and critical-point properties

Substance	Formula	Molar mass kg/kmol	R kJ/(kg · K)*	Temperature K	Pressure MPa	Volume m³/kmol
Ammonia	NH_3	17.03	0.4882	405.5	11.28	0.0724
Argon	Ar	39.948	0.2081	151	4.86	0.0749
Bromine	Br_2	159.808	0.0520	584	10.34	0.1355
Carbon dioxide	CO_2	44.01	0.1889	304.2	7.39	0.0943
Carbon monoxide	CO	28.011	0.2968	133	3.50	0.0930
Chlorine	Cl_2	70.906	0.1173	417	7.71	0.1242
Deuterium (normal)	D_2	4.00	2.0785	38.4	1.66	—
Helium	He	4.003	2.0769	5.3	0.23	0.0578
Hydrogen (normal)	H_2	2.016	4.1240	33.3	1.30	0.0649
Krypton	Kr	83.80	0.09921	209.4	5.50	0.0924
Neon	Ne	20.183	0.4119	44.5	2.73	0.0417
Nitrogen	N_2	28.013	0.2968	126.2	3.39	0.0899
Nitrous oxide	N_2O	44.013	0.1889	309.7	7.27	0.0961
Oxygen	O_2	31.999	0.2598	154.8	5.08	0.0780
Sulfur dioxide	SO_2	64.063	0.1298	430.7	7.88	0.1217
Water	H_2O	18.015	0.4615	647.3	22.09	0.0568
Xenon	Xe	131.30	0.06332	289.8	5.88	0.1186
Benzene	C_6H_6	78.115	0.1064	562	4.92	0.2603
n-Butane	C_4H_{10}	58.124	0.1430	425.2	3.80	0.2547
Carbon tetrachloride	CCl_4	153.82	0.05405	556.4	4.56	0.2759
Chloroform	$CHCl_3$	119.38	0.06964	536.6	5.47	0.2403
Dichlorodifluoromethane (R-12)	CCl_2F_2	120.91	0.06876	384.7	4.01	0.2179
Dichlorofluoromethane	$CHCl_2F$	102.92	0.08078	451.7	5.17	0.1973
Ethane	C_2H_6	30.070	0.2765	305.5	4.88	0.1480
Ethyl alcohol	C_2H_5OH	46.07	0.1805	516	6.38	0.1673
Ethylene	C_2H_4	28.054	0.2964	282.4	5.12	0.1242
n-Hexane	C_6H_{14}	86.178	0.09647	507.9	3.03	0.3677
Methane	CH_4	16.043	0.5182	191.1	4.64	0.0993
Methyl alcohol	CH_3OH	32.042	0.2595	513.2	7.95	0.1180
Methyl chloride	CH_3Cl	50.488	0.1647	416.3	6.68	0.1430
Propane	C_3H_8	44.097	0.1885	370	4.26	0.1998
Propene	C_3H_6	42.081	0.1976	365	4.62	0.1810
Propyne	C_3H_4	40.065	0.2075	401	5.35	—
Trichlorofluoromethane	CCl_3F	137.37	0.06052	471.2	4.38	0.2478
Air	—	28.97	0.2870	—	—	—

*The unit kJ/(kg · K) is equivalent to kPa · m³/(kg · K). The gas constant is calculated from $R = R_u/M$, where $R_u = 8.314$ kJ/(kmol · K) and M is the molar mass.

Source: Gordon J. Van Wylen and Richard E. Sonntag, *Fundamentals of Classical Thermodynamics,* English/SI Version, 3d ed., Wiley, New York, 1986, p. 685, table A.6SI. Originally published in K. A. Kobe and R. E. Lynn, Jr., *Chemical Review,* vol. 52, pp. 117–236, 1953.

R134a
(CF_3CH_2F)

$T_{cr} = 374$ K
$P_{cr} = 4.07$ mPa

Ideal-gas specific heats of various common gases
(*a*) At 300 K

Gas	Formula	Gas constant R kJ/(kg · K)	C_{p0} kJ/(kg · K)	C_{v0} kJ/(kg · K)	k
Air	—	0.2870	1.005	0.718	1.400
Argon	Ar	0.2081	0.5203	0.3122	1.667
Butane	C_4H_{10}	0.1433	1.7164	1.5734	1.091
Carbon dioxide	CO_2	0.1889	0.846	0.657	1.289
Carbon monoxide	CO	0.2968	1.040	0.744	1.400
Ethane	C_2H_6	0.2765	1.7662	1.4897	1.186
Ethylene	C_2H_4	0.2964	1.5482	1.2518	1.237
Helium	He	2.0769	5.1926	3.1156	1.667
Hydrogen	H_2	4.1240	14.307	10.183	1.405
Methane	CH_4	0.5182	2.2537	1.7354	1.299
Neon	Ne	0.4119	1.0299	0.6179	1.667
Nitrogen	N_2	0.2968	1.039	0.743	1.400
Octane	C_8H_{18}	0.0729	1.7113	1.6385	1.044
Oxygen	O_2	0.2598	0.918	0.658	1.395
Propane	C_3H_8	0.1885	1.6794	1.4909	1.126
Steam	H_2O	0.4615	1.8723	1.4108	1.327

Source: Gordon J. Van Wylen and Richard E. Sonntag, *Fundamentals of Classical Thermodynamics,* English/SI Version, 3d ed., Wiley, New York, 1986, p. 687, table A.8SI.

TABLE A-2
(Continued)

(b) At various temperatures

Temperature K	C_{p_0} kJ/(kg·K)	C_{v_0} kJ/(kg·K)	k	C_{p_0} kJ/(kg·K)	C_{v_0} kJ/(kg·K)	k	C_{p_0} kJ/(kg·K)	C_{v_0} kJ/(kg·K)	k
	Air			**Carbon dioxide, CO_2**			**Carbon monoxide, CO**		
250	1.003	0.716	1.401	0.791	0.602	1.314	1.039	0.743	1.400
300	1.005	0.718	1.400	0.846	0.657	1.288	1.040	0.744	1.399
350	1.008	0.721	1.398	0.895	0.706	1.268	1.043	0.746	1.398
400	1.013	0.726	1.395	0.939	0.750	1.252	1.047	0.751	1.395
450	1.020	0.733	1.391	0.978	0.790	1.239	1.054	0.757	1.392
500	1.029	0.742	1.387	1.014	0.825	1.229	1.063	0.767	1.387
550	1.040	0.753	1.381	1.046	0.857	1.220	1.075	0.778	1.382
600	1.051	0.764	1.376	1.075	0.886	1.213	1.087	0.790	1.376
650	1.063	0.776	1.370	1.102	0.913	1.207	1.100	0.803	1.370
700	1.075	0.788	1.364	1.126	0.937	1.202	1.113	0.816	1.364
750	1.087	0.800	1.359	1.148	0.959	1.197	1.126	0.829	1.358
800	1.099	0.812	1.354	1.169	0.980	1.193	1.139	0.842	1.353
900	1.121	0.834	1.344	1.204	1.015	1.186	1.163	0.866	1.343
1000	1.142	0.855	1.336	1.234	1.045	1.181	1.185	0.888	1.335
	Hydrogen, H_2			**Nitrogen, N_2**			**Oxygen, O_2**		
250	14.051	9.927	1.416	1.039	0.742	1.400	0.913	0.653	1.398
300	14.307	10.183	1.405	1.039	0.743	1.400	0.918	0.658	1.395
350	14.427	10.302	1.400	1.041	0.744	1.399	0.928	0.668	1.389
400	14.476	10.352	1.398	1.044	0.747	1.397	0.941	0.681	1.382
450	14.501	10.377	1.398	1.049	0.752	1.395	0.956	0.696	1.373
500	14.513	10.389	1.397	1.056	0.759	1.391	0.972	0.712	1.365
550	14.530	10.405	1.396	1.065	0.768	1.387	0.988	0.728	1.358
600	14.546	10.422	1.396	1.075	0.778	1.382	1.003	0.743	1.350
650	14.571	10.447	1.395	1.086	0.789	1.376	1.017	0.758	1.343
700	14.604	10.480	1.394	1.098	0.801	1.371	1.031	0.771	1.337
750	14.645	10.521	1.392	1.110	0.813	1.365	1.043	0.783	1.332
800	14.695	10.570	1.390	1.121	0.825	1.360	1.054	0.794	1.327
900	14.822	10.698	1.385	1.145	0.849	1.349	1.074	0.814	1.319
1000	14.983	10.859	1.380	1.167	0.870	1.341	1.090	0.830	1.313

Source: Kenneth Wark, *Thermodynamics*, 4th ed., McGraw-Hill, New York, 1983, p. 783, table A-4M. Originally published in *Tables of Thermal Properties of Gases,* NBS Circ. 564, 1955.

(*c*) As a function of temperature

$$\bar{C}_{p0} = a + bT + cT^2 + dT^3$$
[T in K, $\bar{C}_{p0}$ in kJ/(kmol · K)]

Substance	Formula	a	b		c		d		Temperature range K	% error Max.	Avg.
Nitrogen	N_2	28.90	−0.1571	$\times 10^{-2}$	0.8081	$\times 10^{-5}$	−2.873	$\times 10^{-9}$	273–1800	0.59	0.34
Oxygen	O_2	25.48	1.520	$\times 10^{-2}$	−0.7155	$\times 10^{-5}$	1.312	$\times 10^{-9}$	273–1800	1.19	0.28
Air		28.11	0.1967	$\times 10^{-2}$	0.4802	$\times 10^{-5}$	−1.966	$\times 10^{-9}$	273–1800	0.72	0.33
Hydrogen	H_2	29.11	−0.1916	$\times 10^{-2}$	0.4003	$\times 10^{-5}$	−0.8704	$\times 10^{-9}$	273–1800	1.01	0.26
Carbon monoxide	CO	28.16	0.1675	$\times 10^{-2}$	0.5372	$\times 10^{-5}$	−2.222	$\times 10^{-9}$	273–1800	0.89	0.37
Carbon dioxide	CO_2	22.26	5.981	$\times 10^{-2}$	−3.501	$\times 10^{-5}$	7.469	$\times 10^{-9}$	273–1800	0.67	0.22
Water vapor	H_2O	32.24	0.1923	$\times 10^{-2}$	1.055	$\times 10^{-5}$	−3.595	$\times 10^{-9}$	273–1800	0.53	0.24
Nitric oxide	NO	29.34	−0.09395	$\times 10^{-2}$	0.9747	$\times 10^{-5}$	−4.187	$\times 10^{-9}$	273–1500	0.97	0.36
Nitrous oxide	N_2O	24.11	5.8632	$\times 10^{-2}$	−3.562	$\times 10^{-5}$	10.58	$\times 10^{-9}$	273–1500	0.59	0.26
Nitrogen dioxide	NO_2	22.9	5.715	$\times 10^{-2}$	−3.52	$\times 10^{-5}$	7.87	$\times 10^{-9}$	273–1500	0.46	0.18
Ammonia	NH_3	27.568	2.5630	$\times 10^{-2}$	0.99072	$\times 10^{-5}$	−6.6909	$\times 10^{-9}$	273–1500	0.91	0.36
Sulfur	S_2	27.21	2.218	$\times 10^{-2}$	−1.628	$\times 10^{-5}$	3.986	$\times 10^{-9}$	273–1800	0.99	0.38
Sulfur dioxide	SO_2	25.78	5.795	$\times 10^{-2}$	−3.812	$\times 10^{-5}$	8.612	$\times 10^{-9}$	273–1800	0.45	0.24
Sulfur trioxide	SO_3	16.40	14.58	$\times 10^{-2}$	−11.20	$\times 10^{-5}$	32.42	$\times 10^{-9}$	273–1300	0.29	0.13
Acetylene	C_2H_2	21.8	9.2143	$\times 10^{-2}$	−6.527	$\times 10^{-5}$	18.21	$\times 10^{-9}$	273–1500	1.46	0.59
Benzene	C_6H_6	−36.22	48.475	$\times 10^{-2}$	−31.57	$\times 10^{-5}$	77.62	$\times 10^{-9}$	273–1500	0.34	0.20
Methanol	CH_4O	19.0	9.152	$\times 10^{-2}$	−1.22	$\times 10^{-5}$	−8.039	$\times 10^{-9}$	273–1000	0.18	0.08
Ethanol	C_2H_6O	19.9	20.96	$\times 10^{-2}$	−10.38	$\times 10^{-5}$	20.05	$\times 10^{-9}$	273–1500	0.40	0.22
Hydrogen chloride	HCl	30.33	−0.7620	$\times 10^{-2}$	1.327	$\times 10^{-5}$	−4.338	$\times 10^{-9}$	273–1500	0.22	0.08
Methane	CH_4	19.89	5.024	$\times 10^{-2}$	1.269	$\times 10^{-5}$	−11.01	$\times 10^{-9}$	273–1500	1.33	0.57
Ethane	C_2H_6	6.900	17.27	$\times 10^{-2}$	−6.406	$\times 10^{-5}$	7.285	$\times 10^{-9}$	273–1500	0.83	0.28
Propane	C_3H_8	−4.04	30.48	$\times 10^{-2}$	−15.72	$\times 10^{-5}$	31.74	$\times 10^{-9}$	273–1500	0.40	0.12
n-Butane	C_4H_{10}	3.96	37.15	$\times 10^{-2}$	−18.34	$\times 10^{-5}$	35.00	$\times 10^{-9}$	273–1500	0.54	0.24
i-Butane	C_4H_{10}	−7.913	41.60	$\times 10^{-2}$	−23.01	$\times 10^{-5}$	49.91	$\times 10^{-9}$	273–1500	0.25	0.13
n-Pentane	C_5H_{12}	6.774	45.43	$\times 10^{-2}$	−22.46	$\times 10^{-5}$	42.29	$\times 10^{-9}$	273–1500	0.56	0.21
n-Hexane	C_6H_{14}	6.938	55.22	$\times 10^{-2}$	−28.65	$\times 10^{-5}$	57.69	$\times 10^{-9}$	273–1500	0.72	0.20
Ethylene	C_2H_4	3.95	15.64	$\times 10^{-2}$	−8.344	$\times 10^{-5}$	17.67	$\times 10^{-9}$	273–1500	0.54	0.13
Propylene	C_3H_6	3.15	23.83	$\times 10^{-2}$	−12.18	$\times 10^{-5}$	24.62	$\times 10^{-9}$	273–1500	0.73	0.17

Source: B. G. Kyle, *Chemical and Process Thermodynamics*, Prentice-Hall, Englewood Cliffs, N.J., 1984. Used with permission.

TABLE A-3
Specific heats of common solids and liquids
(a) At 25°C

Solid	C_p kJ/(kg · K)	ρ kg/m³	Liquid	C_p kJ/(kg · K)	ρ kg/m³
Aluminum	0.900	2,700	Ammonia	4.800	602
Copper	0.386	8,900	Ethanol	2.456	783
Granite	1.017	2,700	Refrigerant-12	0.977	1,310
Graphite	0.711	2,500	Mercury	0.139	13,560
Iron	0.450	7,840	Methanol	2.550	787
Lead	0.128	11,310	Oil (light)	1.800	910
Rubber (soft)	1.840	1,100	Water	4.184	997
Silver	0.235	10,470			
Tin	0.217	5,730			
Wood (most)	1.760	350–700			

Source: Gordon J. Van Wylen and Richard E. Sonntag, *Fundamentals of Classical Thermodynamics,* English/SI Version, 3d ed., Wiley, New York, 1986, p. 686, table A-7SI.

(*b*) At various temperatures

Solids

Substance	Temp.	C_p kJ/(kg · K)	Substance	Temp.	C_p kJ/(kg · K)
Ice	200 K	1.56	Silver	20°C	0.233
	220 K	1.71		200°C	0.243
	240 K	1.86	Lead	−173°C	0.118
	260 K	2.01		−50°C	0.126
	270 K	2.08		27°C	0.129
	273 K	2.11		100°C	0.131
Aluminum	200 K	0.797		200°C	0.136
	250 K	0.859	Copper	−173°C	0.254
	300 K	0.902		−100°C	0.342
	350 K	0.929		−50°C	0.367
	400 K	0.949		0°C	0.381
	450 K	0.973		27°C	0.386
	500 K	0.997		100°C	0.393
Iron	20°C	0.448		200°C	0.403

Liquids

Substance	State	C_p kJ/(kg · K)	Substance	State	C_p kJ/(kg · K)
Water	1 atm, 273 K	4.217	Benzene	1 atm, 15°C	1.80
	1 atm, 280 K	4.198		1 atm, 65°C	1.92
	1 atm, 300 K	4.179	Glycerin	1 atm, 10°C	2.32
	1 atm, 320 K	4.180		1 atm, 50°C	2.58
	1 atm, 340 K	4.188	Mercury	1 atm, 10°C	0.138
	1 atm, 360 K	4.203		1 atm, 315°C	0.134
	1 atm, 373 K	4.218	Sodium	1 atm, 95°C	1.38
Ammonia	Sat., −20°C	4.52		1 atm, 540°C	1.26
	Sat., 50°C	5.10	Propane	1 atm, 0°C	2.41
Refrigerant-12	Sat., −40°C	0.883	Bismuth	1 atm, 425°C	0.144
	Sat., −20°C	0.908		1 atm, 760°C	0.164
	Sat., 50°C	1.02	Ethyl alcohol	1 atm, 25°C	2.43

Source: Adapted from Kenneth Wark, *Thermodynamics,* 4th ed., McGraw-Hill, New York, 1983, p. 813, table A-19M.

TABLE A-4
Saturated water–Temperature table

Temp. °C T	Sat. press. kPa P_{sat}	Specific volume m³/kg		Internal energy kJ/kg			Enthalpy kJ/kg			Entropy kJ/(kg·K)		
		Sat. liquid v_f	Sat. vapor v_g	Sat. liquid u_f	Evap. u_{fg}	Sat. vapor u_g	Sat. liquid h_f	Evap. h_{fg}	Sat. vapor h_g	Sat. liquid s_f	Evap. s_{fg}	Sat. vapor s_g
0.01	0.6113	0.001 000	206.14	0.0	2375.3	2375.3	0.01	2501.3	2501.4	0.000	9.1562	9.1562
5	0.8721	0.001 000	147.12	20.97	2361.3	2382.3	20.98	2489.6	2510.6	0.0761	8.9496	9.0257
10	1.2276	0.001 000	106.38	42.00	2347.2	2389.2	42.01	2477.7	2519.8	0.1510	8.7498	8.9008
15	1.7051	0.001 001	77.93	62.99	2333.1	2396.1	62.99	2465.9	2528.9	0.2245	8.5569	8.7814
20	2.339	0.001 002	57.79	83.95	2319.0	2402.9	83.96	2454.1	2538.1	0.2966	8.3706	8.6672
25	3.169	0.001 003	43.36	104.88	2304.9	2409.8	104.89	2442.3	2547.2	0.3674	8.1905	8.5580
30	4.246	0.001 004	32.89	125.78	2290.8	2416.6	125.79	2430.5	2556.3	0.4369	8.0164	8.4533
35	5.628	0.001 006	25.22	146.67	2276.7	2423.4	146.68	2418.6	2565.3	0.5053	7.8478	8.3531
40	7.384	0.001 008	19.52	167.56	2262.6	2430.1	167.57	2406.7	2574.3	0.5725	7.6845	8.2570
45	9.593	0.001 010	15.26	188.44	2248.4	2436.8	188.45	2394.8	2583.2	0.6387	7.5261	8.1648
50	12.349	0.001 012	12.03	209.32	2234.2	2443.5	209.33	2382.7	2592.1	0.7038	7.3725	8.0763
55	15.758	0.001 015	9.568	230.21	2219.9	2450.1	230.23	2370.7	2600.9	0.7679	7.2234	7.9913
60	19.940	0.001 017	7.671	251.11	2205.5	2456.6	251.13	2358.5	2609.6	0.8312	7.0784	7.9096
65	25.03	0.001 020	6.197	272.02	2191.1	2463.1	272.06	2346.2	2618.3	0.8935	6.9375	7.8310
70	31.19	0.001 023	5.042	292.95	2176.6	2469.6	292.98	2333.8	2626.8	0.9549	6.8004	7.7553
75	38.58	0.001 026	4.131	313.90	2162.0	2475.9	313.93	2321.4	2635.3	1.0155	6.6669	7.6824
80	47.39	0.001 029	3.407	334.86	2147.4	2482.2	334.91	2308.8	2643.7	1.0753	6.5369	7.6122
85	57.83	0.001 033	2.828	355.84	2132.6	2488.4	355.90	2296.0	2651.9	1.1343	6.4102	7.5445
90	70.14	0.001 036	2.361	376.85	2117.7	2494.5	376.92	2283.2	2660.1	1.1925	6.2866	7.4791
95	84.55	0.001 040	1.982	397.88	2102.7	2500.6	397.96	2270.2	2668.1	1.2500	6.1659	7.4159

Temp. °C T	Sat. press. MPa P_{sat}	Sat. liquid v_f	Sat. vapor v_g	Sat. liquid u_f	Evap. u_{fg}	Sat. vapor u_g	Sat. liquid h_f	Evap. h_{fg}	Sat. vapor h_g	Sat. liquid s_f	Evap. s_{fg}	Sat. vapor s_g
100	0.101 35	0.001 044	1.6729	418.94	2087.6	2506.5	419.04	2257.0	2676.1	1.3069	6.0480	7.3549
105	0.120 82	0.001 048	1.4194	440.02	2072.3	2512.4	440.15	2243.7	2683.8	1.3630	5.9328	7.2958
110	0.143 27	0.001 052	1.2102	461.14	2057.0	2518.1	461.30	2230.2	2691.5	1.4185	5.8202	7.2387
115	0.169 06	0.001 056	1.0366	482.30	2041.4	2523.7	482.48	2216.5	2699.0	1.4734	5.7100	7.1833
120	0.198 53	0.001 060	0.8919	503.50	2025.8	2529.3	503.71	2202.6	2706.3	1.5276	5.6020	7.1296
125	0.2321	0.001 065	0.7706	524.74	2009.9	2534.6	524.99	2188.5	2713.5	1.5813	5.4962	7.0775
130	0.2701	0.001 070	0.6685	546.02	1993.9	2539.9	546.31	2174.2	2720.5	1.6344	5.3925	7.0269
135	0.3130	0.001 075	0.5822	567.35	1977.7	2545.0	567.69	2159.6	2727.3	1.6870	5.2907	6.9777
140	0.3613	0.001 080	0.5089	588.74	1961.3	2550.0	589.13	2144.7	2733.9	1.7391	5.1908	6.9299
145	0.4154	0.001 085	0.4463	610.18	1944.7	2554.9	610.63	2129.6	2740.3	1.7907	5.0926	6.8833
150	0.4758	0.001 091	0.3928	631.68	1927.9	2559.5	632.20	2114.3	2746.5	1.8418	4.9960	6.8379
155	0.5431	0.001 096	0.3468	653.24	1910.8	2564.1	653.84	2098.6	2752.4	1.8925	4.9010	6.7935
160	0.6178	0.001 102	0.3071	674.87	1893.5	2568.4	675.55	2082.6	2758.1	1.9427	4.8075	6.7502
165	0.7005	0.001 108	0.2727	696.56	1876.0	2572.5	697.34	2066.2	2763.5	1.9925	4.7153	6.7078
170	0.7917	0.001 114	0.2428	718.33	1858.1	2576.5	719.21	2049.5	2768.7	2.0419	4.6244	6.6663
175	0.8920	0.001 121	0.2168	740.17	1840.0	2580.2	741.17	2032.4	2773.6	2.0909	4.5347	6.6256
180	1.0021	0.001 127	0.194 05	762.09	1821.6	2583.7	763.22	2015.0	2778.2	2.1396	4.4461	6.5857
185	1.1227	0.001 134	0.174 09	784.10	1802.9	2587.0	785.37	1997.1	2782.4	2.1879	4.3586	6.5465
190	1.2544	0.001 141	0.156 54	806.19	1783.8	2590.0	807.62	1978.8	2786.4	2.2359	4.2720	6.5079
195	1.3978	0.001 149	0.141 05	828.37	1764.4	2592.8	829.98	1960.0	2790.0	2.2835	4.1863	6.4698

Temp. °C T	Sat. press. MPa P_{sat}	Specific volume m³/kg		Internal energy kJ/kg			Enthalpy kJ/kg			Entropy kJ/(kg · K)		
		Sat. liquid v_f	Sat. vapor v_g	Sat. liquid u_f	Evap. u_{fg}	Sat. vapor u_g	Sat. liquid h_f	Evap. h_{fg}	Sat. vapor h_g	Sat. liquid s_f	Evap. s_{fg}	Sat. vapor s_g
200	1.5538	0.001 157	0.127 36	850.65	1744.7	2595.3	852.45	1940.7	2793.2	2.3309	4.1014	6.4323
205	1.7230	0.001 164	0.115 21	873.04	1724.5	2597.5	875.04	1921.0	2796.0	2.3780	4.0172	6.3952
210	1.9062	0.001 173	0.104 41	895.53	1703.9	2599.5	897.76	1900.7	2798.5	2.4248	3.9337	6.3585
215	2.104	0.001 181	0.094 79	918.14	1682.9	2601.1	920.62	1879.9	2800.5	2.4714	3.8507	6.3221
220	2.318	0.001 190	0.086 19	940.87	1661.5	2602.4	943.62	1858.5	2802.1	2.5178	3.7683	6.2861
225	2.548	0.001 199	0.078 49	963.73	1639.6	2603.3	966.78	1836.5	2803.3	2.5639	3.6863	6.2503
230	2.795	0.001 209	0.071 58	986.74	1617.2	2603.9	990.12	1813.8	2804.0	2.6099	3.6047	6.2146
235	3.060	0.001 219	0.065 37	1009.89	1594.2	2604.1	1013.62	1790.5	2804.2	2.6558	3.5233	6.1791
240	3.344	0.001 229	0.059 76	1033.21	1570.8	2604.0	1037.32	1766.5	2803.8	2.7015	3.4422	6.1437
245	3.648	0.001 240	0.054 71	1056.71	1546.7	2603.4	1061.23	1741.7	2803.0	2.7472	3.3612	6.1083
250	3.973	0.001 251	0.050 13	1080.39	1522.0	2602.4	1085.36	1716.2	2801.5	2.7927	3.2802	6.0730
255	4.319	0.001 263	0.045 98	1104.28	1596.7	2600.9	1109.73	1689.8	2799.5	2.8383	3.1992	6.0375
260	4.688	0.001 276	0.042 21	1128.39	1470.6	2599.0	1134.37	1662.5	2796.9	2.8838	3.1181	6.0019
265	5.081	0.001 289	0.038 77	1152.74	1443.9	2596.6	1159.28	1634.4	2793.6	2.9294	3.0368	5.9662
270	5.499	0.001 302	0.035 64	1177.36	1416.3	2593.7	1184.51	1605.2	2789.7	2.9751	2.9551	5.9301
275	5.942	0.001 317	0.032 79	1202.25	1387.9	2590.2	1210.07	1574.9	2785.0	3.0208	2.8730	5.8938
280	6.412	0.001 332	0.030 17	1227.46	1358.7	2586.1	1235.99	1543.6	2779.6	3.0668	2.7903	5.8571
285	6.909	0.001 348	0.027 77	1253.00	1328.4	2581.4	1262.31	1511.0	2773.3	3.1130	2.7070	5.8199
290	7.436	0.001 366	0.025 57	1278.92	1297.1	2576.0	1289.07	1477.1	2766.2	3.1594	2.6227	5.7821
295	7.993	0.001 384	0.023 54	1305.2	1264.7	2569.9	1316.3	1441.8	2758.1	3.2062	2.5375	5.7437
300	8.581	0.001 404	0.021 67	1332.0	1231.0	2563.0	1344.0	1404.9	2749.0	3.2534	2.4511	5.7045
305	9.202	0.001 425	0.019 948	1359.3	1195.9	2555.2	1372.4	1366.4	2738.7	3.3010	2.3633	5.6643
310	9.856	0.001 447	0.018 350	1387.1	1159.4	2546.4	1401.3	1326.0	2727.3	3.3493	2.2737	5.6230
315	10.547	0.001 472	0.016 867	1415.5	1121.1	2536.6	1431.0	1283.5	2714.5	3.3982	2.1821	5.5804
320	11.274	0.001 499	0.015 488	1444.6	1080.9	2525.5	1461.5	1238.6	2700.1	3.4480	2.0882	5.5362
330	12.845	0.001 561	0.012 996	1505.3	993.7	2498.9	1525.3	1140.6	2665.9	3.5507	1.8909	5.4417
340	14.586	0.001 638	0.010 797	1570.3	894.3	2464.6	1594.2	1027.9	2622.0	3.6594	1.6763	5.3357
350	16.513	0.001 740	0.008 813	1641.9	776.6	2418.4	1670.6	893.4	2563.9	3.7777	1.4335	5.2112
360	18.651	0.001 893	0.006 945	1725.2	626.3	2351.5	1760.5	720.3	2481.0	3.9147	1.1379	5.0526
370	21.03	0.002 213	0.004 925	1844.0	384.5	2228.5	1890.5	441.6	2332.1	4.1106	0.6865	4.7971
374.14	22.09	0.003 155	0.003 155	2029.6	0	2029.6	2099.3	0	2099.3	4.4298	0	4.4298

Source: Tables A-4 through A-8 are adapted from Gordon J. Van Wylen and Richard E. Sonntag, *Fundamentals of Classical Thermodynamics,* English/SI Version, 3d ed., Wiley, New York, 1986, pp. 635–651. Originally published in Joseph H. Keenan, Frederick G. Keyes, Philip G. Hill, and Joan G. Moore, *Steam Tables*, SI Units, Wiley, New York, 1978.

TABLE A-5
Saturated water–Pressure table

Press. kPa P	Sat. temp. °C T_{sat}	Specific volume m³/kg Sat. liquid v_f	Sat. vapor v_g	Internal energy kJ/kg Sat. liquid u_f	Evap. u_{fg}	Sat. vapor u_g	Enthalpy kJ/kg Sat. liquid h_f	Evap. h_{fg}	Sat. vapor h_g	Entropy kJ/(kg·K) Sat. liquid s_f	Evap. s_{fg}	Sat. vapor s_g
0.6113	0.01	0.001 000	206.14	0.00	2375.3	2375.3	0.01	2501.3	2501.4	0.0000	9.1562	9.1562
1.0	6.98	0.001 000	129.21	29.30	2355.7	2385.0	29.30	2484.9	2514.2	0.1059	8.8697	8.9756
1.5	13.03	0.001 001	87.98	54.71	2338.6	2393.3	54.71	2470.6	2525.3	0.1957	8.6322	8.8279
2.0	17.50	0.001 001	67.00	73.48	2326.0	2399.5	73.48	2460.0	2533.5	0.2607	8.4629	8.7237
2.5	21.08	0.001 002	54.25	88.48	2315.9	2404.4	88.49	2451.6	2540.0	0.3120	8.3311	8.6432
3.0	24.08	0.001 003	45.67	101.04	2307.5	2408.5	101.05	2444.5	2545.5	0.3545	8.2231	8.5776
4.0	28.96	0.001 004	34.80	121.45	2293.7	2415.2	121.46	2432.9	2554.4	0.4226	8.0520	8.4746
5.0	32.88	0.001 005	28.19	137.81	2282.7	2420.5	137.82	2423.7	2561.5	0.4764	7.9187	8.3951
7.5	40.29	0.001 008	19.24	168.78	2261.7	2430.5	168.79	2406.0	2574.8	0.5764	7.6750	8.2515
10	45.81	0.001 010	14.67	191.82	2246.1	2437.9	191.83	2392.8	2584.7	0.6493	7.5009	8.1502
15	53.97	0.001 014	10.02	225.92	2222.8	2448.7	225.94	2373.1	2599.1	0.7549	7.2536	8.0085
20	60.06	0.001 017	7.649	251.38	2205.4	2456.7	251.40	2358.3	2609.7	0.8320	7.0766	7.9085
25	64.97	0.001 020	6.204	271.90	2191.2	2463.1	271.93	2346.3	2618.2	0.8931	6.9383	7.8314
30	69.10	0.001 022	5.229	289.20	2179.2	2468.4	289.23	2336.1	2625.3	0.9439	6.8247	7.7686
40	75.87	0.001 027	3.993	317.53	2159.5	2477.0	317.58	2319.2	2636.8	1.0259	6.6441	7.6700
50	81.33	0.001 030	3.240	340.44	2143.4	2483.9	340.49	2305.4	2645.9	1.0910	6.5029	7.5939
75	91.78	0.001 037	2.217	384.31	2112.4	2496.7	384.39	2278.6	2663.0	1.2130	6.2434	7.4564

Press. MPa P	Sat. temp. °C T_{sat}	Specific volume m³/kg Sat. liquid v_f	Sat. vapor v_g	Internal energy kJ/kg Sat. liquid u_f	Evap. u_{fg}	Sat. vapor u_g	Enthalpy kJ/kg Sat. liquid h_f	Evap. h_{fg}	Sat. vapor h_g	Entropy kJ/(kg·K) Sat. liquid s_f	Evap. s_{fg}	Sat. vapor s_g
0.100	99.63	0.001 043	1.6940	417.36	2088.7	2506.1	417.46	2258.0	2675.5	1.3026	6.0568	7.3594
0.125	105.99	0.001 048	1.3749	444.19	2069.3	2513.5	444.32	2241.0	2685.4	1.3740	5.9104	7.2844
0.150	111.37	0.001 053	1.1593	466.94	2052.7	2519.7	467.11	2226.5	2693.6	1.4336	5.7897	7.2233
0.175	116.06	0.001 057	1.0036	486.80	2038.1	2524.9	486.99	2213.6	2700.6	1.4849	5.6868	7.1717
0.200	120.23	0.001 061	0.8857	504.49	2025.0	2529.5	504.70	2201.9	2706.7	1.5301	5.5970	7.1271
0.225	124.00	0.001 064	0.7933	520.47	2013.1	2533.6	520.72	2191.3	2712.1	1.5706	5.5173	7.0878
0.250	127.44	0.001 067	0.7187	535.10	2002.1	2537.2	535.37	2181.5	2716.9	1.6072	5.4455	7.0527
0.275	130.60	0.001 070	0.6573	548.59	1991.9	2540.5	548.89	2172.4	2721.3	1.6408	5.3801	7.0209
0.300	133.55	0.001 073	0.6058	561.15	1982.4	2543.6	561.47	2163.8	2725.3	1.6718	5.3201	6.9919
0.325	136.30	0.001 076	0.5620	572.90	1973.5	2546.4	573.25	2155.8	2729.0	1.7006	5.2646	6.9652
0.350	138.88	0.001 079	0.5243	583.95	1965.0	2548.9	584.33	2148.1	2732.4	1.7275	5.2130	6.9405
0.375	141.32	0.001 081	0.4914	594.40	1956.9	2551.3	594.81	2140.8	2735.6	1.7528	5.1647	6.9175
0.40	143.63	0.001 084	0.4625	604.31	1949.3	2553.6	604.74	2133.8	2738.6	1.7766	5.1193	6.8959
0.45	147.93	0.001 088	0.4140	622.77	1934.9	2557.6	623.25	2120.7	2743.9	1.8207	5.0359	6.8565
0.50	151.86	0.001 093	0.3749	639.68	1921.6	2561.2	640.23	2108.5	2748.7	1.8607	4.9606	6.8213
0.55	155.48	0.001 097	0.3427	655.32	1909.2	2564.5	665.93	2097.0	2753.0	1.8973	4.8920	6.7893
0.60	158.85	0.001 101	0.3157	669.90	1897.5	2567.4	670.56	2086.3	2756.8	1.9312	4.8288	6.7600
0.65	162.01	0.001 104	0.2927	683.56	1886.5	2570.1	684.28	2076.0	2760.3	1.9627	4.7703	6.7331
0.70	164.97	0.001 108	0.2729	696.44	1876.1	2572.5	697.22	2066.3	2763.5	1.9922	4.7158	6.7080
0.75	167.78	0.001 112	0.2556	708.64	1866.1	2574.7	709.47	2057.0	2766.4	2.0200	4.6647	6.6847
0.80	170.43	0.001 115	0.2404	720.22	1856.6	2576.8	721.11	2048.0	2769.1	2.0462	4.6166	6.6628
0.85	172.96	0.001 118	0.2270	731.27	1847.4	2578.7	732.22	2039.4	2771.6	2.0710	4.5711	6.6421
0.90	175.38	0.001 121	0.2150	741.83	1838.6	2580.5	742.83	2031.1	2773.9	2.0946	4.5280	6.6226
0.95	177.69	0.001 124	0.2042	751.95	1830.2	2582.1	753.02	2023.1	2776.1	2.1172	4.4869	6.6041
1.00	179.91	0.001 127	0.194 44	761.68	1822.0	2583.6	762.81	2015.3	2778.1	2.1387	4.4478	6.5865
1.10	184.09	0.001 133	0.177 53	780.09	1806.3	2586.4	781.34	2000.4	2781.7	2.1792	4.3744	6.5536
1.20	187.99	0.001 139	0.163 33	797.29	1791.5	2588.8	798.65	1986.2	2784.8	2.2166	4.3067	6.5233
1.30	191.64	0.001 144	0.151 25	813.44	1777.5	2591.0	814.93	1972.7	2787.6	2.2515	4.2438	6.4953

Press. MPa P	Sat. temp. °C T_{sat}	Specific volume m³/kg		Internal energy kJ/kg			Enthalpy kJ/kg			Entropy kJ/(kg·K)		
		Sat. liquid v_f	Sat. vapor v_g	Sat. liquid u_f	Evap. u_{fg}	Sat. vapor u_g	Sat. liquid h_f	Evap. h_{fg}	Sat. vapor h_g	Sat. liquid s_f	Evap. s_{fg}	Sat. vapor s_g
1.40	195.07	0.001 149	0.140 84	828.70	1764.1	2592.8	830.30	1959.7	2790.0	2.2842	4.1850	6.4693
1.50	198.32	0.001 154	0.131 77	843.16	1751.3	2594.5	844.89	1947.3	2792.2	2.3150	4.1298	6.4448
1.75	205.76	0.001 166	0.113 49	876.46	1721.4	2597.8	878.50	1917.9	2796.4	2.3851	4.0044	6.3896
2.00	212.42	0.001 177	0.099 63	906.44	1693.8	2600.3	908.79	1890.7	2799.5	2.4474	3.8935	6.3409
2.25	218.45	0.001 187	0.088 75	933.83	1668.2	2602.0	936.49	1865.2	2801.7	2.5035	3.7937	6.2972
2.5	223.99	0.001 197	0.079 98	959.11	1644.0	2603.1	962.11	1841.0	2803.1	2.5547	3.7028	6.2575
3.0	233.90	0.001 217	0.066 68	1004.78	1599.3	2604.1	1008.42	1795.7	2804.2	2.6457	3.5412	6.1869
3.5	242.60	0.001 235	0.057 07	1045.43	1558.3	2603.7	1049.75	1753.7	2803.4	2.7253	3.4000	6.1253
4	250.40	0.001 252	0.049 78	1082.31	1520.0	2602.3	1087.31	1714.1	2801.4	2.7964	3.2737	6.0701
5	263.99	0.001 286	0.039 44	1147.81	1449.3	2597.1	1154.23	1640.1	2794.3	2.9202	3.0532	5.9734
6	275.64	0.001 319	0.032 44	1205.44	1384.3	2589.7	1213.35	1571.0	2784.3	3.0267	2.8625	5.8892
7	285.88	0.001 351	0.027 37	1257.55	1323.0	2580.5	1267.00	1505.1	2772.1	3.1211	2.6922	5.8133
8	295.06	0.001 384	0.023 52	1305.57	1264.2	2569.8	1316.64	1441.3	2758.0	3.2068	2.5364	5.7432
9	303.40	0.001 418	0.020 48	1350.51	1207.3	2557.8	1363.26	1378.9	2742.1	3.2858	2.3915	5.6722
10	311.06	0.001 452	0.018 026	1393.04	1151.4	2544.4	1407.56	1317.1	2724.7	3.3596	2.2544	5.6141
11	318.15	0.001 489	0.015 987	1433.7	1096.0	2529.8	1450.1	1255.5	2705.6	3.4295	2.1233	5.5527
12	324.75	0.001 527	0.014 263	1473.0	1040.7	2513.7	1491.3	1193.3	2684.9	3.4962	1.9962	5.4924
13	330.93	0.001 567	0.012 780	1511.1	985.0	2496.1	1531.5	1130.7	2662.2	3.5606	1.8718	5.4323
14	336.75	0.001 611	0.011 485	1548.6	928.2	2476.8	1571.1	1066.5	2637.6	3.6232	1.7485	5.3717
15	342.24	0.001 658	0.010 337	1585.6	869.8	2455.5	1610.5	1000.0	2610.5	3.6848	1.6249	5.3098
16	347.44	0.001 711	0.009 306	1622.7	809.0	2431.7	1650.1	930.6	2580.6	3.7461	1.4994	5.2455
17	352.37	0.001 770	0.008 364	1660.2	744.8	2405.0	1690.3	856.9	2547.2	3.8079	1.3698	5.1777
18	357.06	0.001 840	0.007 489	1698.9	675.4	2374.3	1732.0	777.1	2509.1	3.8715	1.2329	5.1044
19	361.54	0.001 924	0.006 657	1739.9	598.1	2338.1	1776.5	688.0	2464.5	3.9388	1.0839	5.0228
20	365.81	0.002 036	0.005 834	1785.6	507.5	2293.0	1826.3	583.4	2409.7	4.0139	0.9130	4.9269
21	369.89	0.002 207	0.004 952	1842.1	388.5	2230.6	1888.4	446.2	2334.6	4.1075	0.6938	4.8013
22	373.80	0.002 742	0.003 568	1961.9	125.2	2087.1	2022.2	143.4	2165.6	4.3110	0.2216	4.5327
22.09	374.14	0.003 155	0.003 155	2029.6	0	2029.6	2099.3	0	2099.3	4.4298	0	4.4298

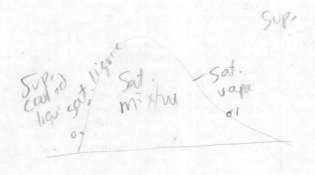

TABLE A-6
Superheated water

T °C	v m³/kg	u kJ/kg	h kJ/kg	s kJ/(kg·K)	v m³/kg	u kJ/kg	h kJ/kg	s kJ/(kg·K)	v m³/kg	u kJ/kg	h kJ/kg	s kJ/(kg·K)
	P = 0.01 MPa (45.81°C)*				P = 0.05 MPa (81.33°C)				P = 0.10 MPa (99.63°C)			
Sat.†	14.674	2437.9	2584.7	8.1502	3.240	2483.9	2645.9	7.5939	1.6940	2506.1	2675.5	7.3594
50	14.869	2443.9	2592.6	8.1749								
100	17.196	2515.5	2687.5	8.4479	3.418	2511.6	2682.5	7.6947	1.6958	2506.7	2676.2	7.3614
150	19.512	2587.9	2783.0	8.6882	3.889	2585.6	2780.1	7.9401	1.9364	2582.8	2776.4	7.6134
200	21.825	2661.3	2879.5	8.9038	4.356	2659.9	2877.7	8.1580	2.172	2658.1	2875.3	7.8343
250	24.136	2736.0	2977.3	9.1002	4.820	2735.0	2976.0	8.3556	2.406	2733.7	2974.3	8.0333
300	26.445	2812.1	3076.5	9.2813	5.284	2811.3	3075.5	8.5373	2.639	2810.4	3074.3	8.2158
400	31.063	2968.9	3279.6	9.6077	6.209	2968.5	3278.9	8.8642	3.103	2967.9	3278.2	8.5435
500	35.679	3132.3	3489.1	9.8978	7.134	3132.0	3488.7	9.1546	3.565	3131.6	3488.1	8.8342
600	40.295	3302.5	3705.4	10.1608	8.057	3302.2	3705.1	9.4178	4.028	3301.9	3704.4	9.0976
700	44.911	3479.6	3928.7	10.4028	8.981	3479.4	3928.5	9.6599	4.490	3479.2	3928.2	9.3398
800	49.526	3663.8	4159.0	10.6281	9.904	3663.6	4158.9	9.8852	4.952	3663.5	4158.6	9.5652
900	54.141	3855.0	4396.4	10.8396	10.828	3854.9	4396.3	10.0967	5.414	3854.8	4396.1	9.7767
1000	58.757	4053.0	4640.6	11.0393	11.751	4052.9	4640.5	10.2964	5.875	4052.8	4640.3	9.9764
1100	63.372	4257.5	4891.2	11.2287	12.674	4257.4	4891.1	10.4859	6.337	4257.3	4891.0	10.1659
1200	67.987	4467.9	5147.8	11.4091	13.597	4467.8	5147.7	10.6662	6.799	4467.7	5147.6	10.3463
1300	72.602	4683.7	5409.7	11.5811	14.521	4683.6	5409.6	10.8382	7.260	4683.5	5409.5	10.5183
	P = 0.20 MPa (120.23°C)				P = 0.30 MPa (133.55°C)				P = 0.40 MPa (143.63°C)			
Sat.	0.8857	2529.5	2706.7	7.1272	0.6058	2543.6	2725.3	6.9919	0.4625	2553.6	2738.6	6.8959
150	0.9596	2576.9	2768.8	7.2795	0.6339	2570.8	2761.0	7.0778	0.4708	2564.5	2752.8	6.9299
200	1.0803	2654.4	2870.5	7.5066	0.7163	2650.7	2865.6	7.3115	0.5342	2646.8	2860.5	7.1706
250	1.1988	2731.2	2971.0	7.7086	0.7964	2728.7	2967.6	7.5166	0.5951	2726.1	2964.2	7.3789
300	1.3162	2808.6	3071.8	7.8926	0.8753	2806.7	3069.3	7.7022	0.6548	2804.8	3066.8	7.5662
400	1.5493	2966.7	3276.6	8.2218	1.0315	2965.6	3275.0	8.0330	0.7726	2964.4	3273.4	7.8985
500	1.7814	3130.8	3487.1	8.5133	1.1867	3130.0	3486.0	8.3251	0.8893	3129.2	3484.9	8.1913
600	2.013	3301.4	3704.0	8.7770	1.3414	3300.8	3703.2	8.5892	1.0055	3300.2	3702.4	8.4558
700	2.244	3478.8	3927.6	9.0194	1.4957	3478.4	3927.1	8.8319	1.1215	3477.9	3926.5	8.6987
800	2.475	3663.1	4158.2	9.2449	1.6499	3662.9	4157.8	9.0576	1.2372	3662.4	4157.3	8.9244
900	2.705	3854.5	4395.8	9.4566	1.8041	3854.2	4395.4	9.2692	1.3529	3853.9	4395.1	9.1362
1000	2.937	4052.5	4640.0	9.6563	1.9581	4052.3	4639.7	9.4690	1.4685	4052.0	4639.4	9.3360
1100	3.168	4257.0	4890.7	9.8458	2.1121	4256.8	4890.4	9.6585	1.5840	4256.5	4890.2	9.5256
1200	3.399	4467.5	5147.5	10.0262	2.2661	4467.2	5147.1	9.8389	1.6996	4467.0	5146.8	9.7060
1300	3.630	4683.2	5409.3	10.1982	2.4201	4683.0	5409.0	10.0110	1.8151	4682.8	5408.8	9.8780
	P = 0.50 MPa (151.86°C)				P = 0.60 MPa (158.85°C)				P = 0.80 MPa (170.43°C)			
Sat.	0.3749	2561.2	2748.7	6.8213	0.3157	2567.4	2756.8	6.7600	0.2404	2576.8	2769.1	6.6628
200	0.4249	2642.9	2855.4	7.0592	0.3520	2638.9	2850.1	6.9665	0.2608	2630.6	2839.3	6.8158
250	0.4744	2723.5	2960.7	7.2709	0.3938	2720.9	2957.2	7.1816	0.2931	2715.5	2950.0	7.0384
300	0.5226	2802.9	3064.2	7.4599	0.4344	2801.0	3061.6	7.3724	0.3241	2797.2	3056.5	7.2328
350	0.5701	2882.6	3167.7	7.6329	0.4742	2881.2	3165.7	7.5464	0.3544	2878.2	3161.7	7.4089
400	0.6173	2963.2	3271.9	7.7938	0.5137	2962.1	3270.3	7.7079	0.3843	2959.7	3267.1	7.5716
500	0.7109	3128.4	3483.9	8.0873	0.5920	3127.6	3482.8	8.0021	0.4433	3126.0	3480.6	7.8673
600	0.8041	3299.6	3701.7	7.3522	0.6697	3299.1	3700.9	8.2674	0.5018	3297.9	3699.4	8.1333
700	0.8969	3477.5	3925.9	8.5952	0.7472	3477.0	3925.3	8.5107	0.5601	3476.2	3924.2	8.3770
800	0.9896	3662.1	4156.9	8.8211	0.8245	3661.8	4156.5	8.7367	0.6181	3661.1	4155.6	8.6033
900	1.0822	3853.6	4394.7	9.0329	0.9017	3853.4	4394.4	8.9486	0.6761	3852.8	4393.7	8.8153
1000	1.1747	4051.8	4639.1	9.2328	0.9788	4051.5	4638.8	9.1485	0.7340	4051.0	4638.2	9.0153
1100	1.2672	4256.3	4889.9	9.4224	1.0559	4256.1	4889.6	9.3381	0.7919	4255.6	4889.1	9.2050
1200	1.3596	4466.8	5146.6	9.6029	1.1330	4466.5	5146.3	9.5185	0.8497	4466.1	5145.9	9.3855
1300	1.4521	4682.5	5408.6	9.7749	1.2101	4682.3	5408.3	9.6906	0.9076	4681.8	5407.9	9.5575

*The temperature in parentheses is the saturation temperature at the specified pressure.
†Properties of saturated vapor at the specified pressure.

T °C	v m³/kg	u kJ/kg	h kJ/kg	s kJ/(kg·K)	v m³/kg	u kJ/kg	h kJ/kg	s kJ/(kg·K)	v m³/kg	u kJ/kg	h kJ/kg	s kJ/(kg·K)
	P = 1.00 MPa (179.91°C)				*P* = 1.20 MPa (187.99°C)				*P* = 1.40 MPa (195.07°C)			
Sat.	0.194 44	2583.6	2778.1	6.5865	0.163 33	2588.8	2784.8	6.5233	0.140 84	2592.8	2790.0	6.4693
200	0.2060	2621.9	2827.9	6.6940	0.169 30	2612.8	2815.9	6.5898	0.143 02	2603.1	2803.3	6.4975
250	0.2327	2709.9	2942.6	6.9247	0.192 34	2704.2	2935.0	6.8294	0.163 50	2698.3	2927.2	6.7467
300	0.2579	2793.2	3051.2	7.1229	0.2138	2789.2	3045.8	7.0317	0.182 28	2785.2	3040.4	6.9534
350	0.2825	2875.2	3157.7	7.3011	0.2345	2872.2	3153.6	7.2121	0.2003	2869.2	3149.5	7.1360
400	0.3066	2957.3	3263.9	7.4651	0.2548	2954.9	3260.7	7.3774	0.2178	2952.5	3257.5	7.3026
500	0.3541	3124.4	3478.5	7.7622	0.2946	3122.8	3476.3	7.6759	0.2521	3121.1	3474.1	7.6027
600	0.4011	3296.8	3697.9	8.0290	0.3339	3295.6	3696.3	7.9435	0.2860	3294.4	3694.8	7.8710
700	0.4478	3475.3	3923.1	8.2731	0.3729	3474.4	3922.0	8.1881	0.3195	3473.6	3920.8	8.1160
800	0.4943	3660.4	4154.7	8.4996	0.4118	3659.7	4153.8	8.4148	0.3528	3659.0	4153.0	8.3431
900	0.5407	3852.2	4392.9	8.7118	0.4505	3851.6	4392.2	8.6272	0.3861	3851.1	4391.5	8.5556
1000	0.5871	4050.5	4637.6	8.9119	0.4892	4050.0	4637.0	8.8274	0.4192	4049.5	4636.4	8.7559
1100	0.6335	4255.1	4888.6	9.1017	0.5278	4254.6	4888.0	9.0172	0.4524	4254.1	4887.5	8.9457
1200	0.6798	4465.6	5145.4	9.2822	0.5665	4465.1	5144.9	9.1977	0.4855	4464.7	5144.4	9.1262
1300	0.7261	4681.3	5407.4	9.4543	0.6051	4680.9	5407.0	9.3698	0.5186	4680.4	5406.5	9.2984
	P = 1.60 MPa (201.41°C)				*P* = 1.80 MPa (207.15°C)				*P* = 2.00 MPa (212.42°C)			
Sat.	0.123 80	2596.0	2794.0	6.4218	0.110 42	2598.4	2797.1	6.3794	0.099 63	2600.3	2799.5	6.3409
225	0.132 87	2644.7	2857.3	6.5518	0.116 73	2636.6	2846.7	6.4808	0.103 77	2628.3	2835.8	6.4147
250	0.141 84	2692.3	2919.2	6.6732	0.124 97	2686.0	2911.0	6.6066	0.111 44	2679.6	2902.5	6.5453
300	0.158 62	2781.1	3034.8	6.8844	0.140 21	2776.9	3029.2	6.8226	0.125 47	2772.6	3023.5	6.7664
350	0.174 56	2866.1	3145.4	7.0694	0.154 57	2863.0	3141.2	7.0100	0.138 57	2859.8	3137.0	6.9563
400	0.190 05	2950.1	3254.2	7.2374	0.168 47	2947.7	3250.9	7.1794	0.151 20	2945.2	3247.6	7.1271
500	0.2203	3119.5	3472.0	7.5390	0.195 50	3117.9	3469.8	7.4825	0.175 68	3116.2	3467.6	7.4317
600	0.2500	3293.3	3693.2	7.8080	0.2220	3292.1	3691.7	7.7523	0.199 60	3290.9	3690.1	7.7024
700	0.2794	3472.7	3919.7	8.0535	0.2482	3471.8	3918.5	7.9983	0.2232	3470.9	3917.4	7.9487
800	0.3086	3658.3	4152.1	8.2808	0.2742	3657.6	4151.2	8.2258	0.2467	3657.0	4150.3	8.1765
900	0.3377	3850.5	4390.8	8.4935	0.3001	3849.9	4390.1	8.4386	0.2700	3849.3	4389.4	8.3895
1000	0.3668	4049.0	4635.8	8.6938	0.3260	4048.5	4635.2	8.6391	0.2933	4048.0	4634.6	8.5901
1100	0.3958	4253.7	4887.0	8.8837	0.3518	4253.2	4886.4	8.8290	0.3166	4252.7	4885.9	8.7800
1200	0.4248	4464.2	5143.9	9.0643	0.3776	4463.7	5143.4	9.0096	0.3398	4463.3	5142.9	8.9607
1300	0.4538	4679.9	5406.0	9.2364	0.4034	4679.5	5405.6	9.1818	0.3631	4679.0	5405.1	9.1329
	P = 2.50 MPa (223.99°C)				*P* = 3.00 MPa (233.90°C)				*P* = 3.50 MPa (242.60°C)			
Sat.	0.079 98	2603.1	2803.1	6.2575	0.066 68	2604.1	2804.2	6.1869	0.057 07	2603.7	2803.4	6.1253
225	0.080 27	2605.6	2806.3	6.2639								
250	0.087 00	2662.6	2880.1	6.4085	0.070 58	2644.0	2855.8	6.2872	0.058 72	2623.7	2829.2	6.1749
300	0.098 90	2761.6	3008.8	6.6438	0.081 14	2750.1	2993.5	6.5390	0.068 42	2738.0	2977.5	6.4461
350	0.109 76	2851.9	3126.3	6.8403	0.090 53	2843.7	3115.3	6.7428	0.076 78	2835.3	3104.0	6.6579
400	0.120 10	2939.1	3239.3	7.0148	0.099 36	2932.8	3230.9	6.9212	0.084 53	2926.4	3222.3	6.8405
450	0.130 14	3025.5	3350.8	7.1746	0.107 87	3020.4	3344.0	7.0834	0.091 96	3015.3	3337.2	7.0052
500	0.139 93	3112.1	3462.1	7.3234	0.116 19	3108.0	3456.5	7.2338	0.099 18	3103.0	3450.9	7.1572
600	0.159 30	3288.0	3686.3	7.5960	0.132 43	3285.0	3682.3	7.5085	0.113 24	3282.1	3678.4	7.4339
700	0.178 32	3468.7	3914.5	7.8435	0.148 38	3466.5	3911.7	7.7571	0.126 99	3464.3	3908.8	7.6837
800	0.197 16	3655.3	4148.2	8.0720	0.164 14	3653.5	4145.9	7.9862	0.140 56	3651.8	4143.7	7.9134
900	0.215 90	3847.9	4387.6	8.2853	0.179 80	3846.5	4385.9	8.1999	0.154 02	3845.0	4384.1	8.1276
1000	0.2346	4046.7	4633.1	8.4861	0.195 41	4045.4	4631.6	8.4009	0.167 43	4044.1	4630.1	8.3288
1100	0.2532	4251.5	4884.6	8.6762	0.210 98	4250.3	4883.3	8.5912	0.180 80	4249.2	4881.9	8.5192
1200	0.2718	4462.1	5141.7	8.8569	0.226 52	4460.9	5140.5	8.7720	0.194 15	4459.8	5139.3	8.7000
1300	0.2905	4677.8	5404.0	9.0291	0.242 06	4676.6	5402.8	8.9442	0.207 49	4675.5	5401.7	8.8723

TABLE A-6
(*Continued*)

T °C	v m³/kg	u kJ/kg	h kJ/kg	s kJ/(kg·K)	v m³/kg	u kJ/kg	h kJ/kg	s kJ/(kg·K)	v m³/kg	u kJ/kg	h kJ/kg	s kJ/(kg·K)
	P = 4.0 MPa (250.40°C)				*P* = 4.5 MPa (257.49°C)				*P* = 5.0 MPa (263.99°C)			
Sat.	0.049 78	2602.3	2801.4	6.0701	0.044 06	2600.1	2798.3	6.0198	0.039 44	2597.1	2794.3	5.9734
275	0.054 57	2667.9	2886.2	6.2285	0.047 30	2650.3	2863.2	6.1401	0.041 41	2631.3	2838.3	6.0544
300	0.058 84	2725.3	2960.7	6.3615	0.051 35	2712.0	2943.1	6.2828	0.045 32	2698.0	2924.5	6.2084
350	0.066 45	2826.7	3092.5	6.5821	0.058 40	2817.8	3080.6	6.5131	0.051 94	2808.7	3068.4	6.4493
400	0.073 41	2919.9	3213.6	6.7690	0.064 75	2913.3	3204.7	6.7047	0.057 81	2906.6	3195.7	6.6459
450	0.080 02	3010.2	3330.3	6.9363	0.070 74	3005.0	3323.3	6.8746	0.063 30	2999.7	3316.2	6.8186
500	0.086 43	3099.5	3445.3	7.0901	0.076 51	3095.3	3439.6	7.0301	0.068 57	3091.0	3433.8	6.9759
600	0.098 85	3279.1	3674.4	7.3688	0.087 65	3276.0	3670.5	7.3110	0.078 69	3273.0	3666.5	7.2589
700	0.110 95	3462.1	3905.9	7.6198	0.098 47	3459.9	3903.0	7.5631	0.088 49	3457.6	3900.1	7.5122
800	0.122 87	3650.0	4141.5	7.8502	0.109 11	3648.3	4139.3	7.7942	0.098 11	3646.6	4137.1	7.7440
900	0.134 69	3843.6	4382.3	8.0647	0.119 65	3842.2	4380.6	8.0091	0.107 62	3840.7	4378.8	7.9593
1000	0.146 45	4042.9	4628.7	8.2662	0.130 13	4041.6	4627.2	8.2108	0.117 07	4040.4	4625.7	8.1612
1100	0.158 17	4248.0	4880.6	8.4567	0.140 56	4246.8	4879.3	8.4015	0.126 48	4245.6	4878.0	8.3520
1200	0.169 87	4458.6	5138.1	8.6376	0.150 98	4457.5	5136.9	8.5825	0.135 87	4456.3	5135.7	8.5331
1300	0.181 56	4674.3	5400.5	8.8100	0.161 39	4673.1	5399.4	8.7549	0.145 26	4672.0	5398.2	8.7055
	P = 6.0 MPa (275.64°C)				*P* = 7.0 MPa (285.88°C)				*P* = 8.0 MPa (295.06°C)			
Sat.	0.032 44	2589.7	2784.3	5.8892	0.027 37	2580.5	2772.1	5.8133	0.023 52	2569.8	2758.0	5.7432
300	0.036 16	2667.2	2884.2	6.0674	0.029 47	2632.2	2838.4	5.9305	0.024 26	2590.9	2785.0	5.7906
350	0.042 23	2789.6	3043.0	6.3335	0.035 24	2769.4	3016.0	6.2283	0.029 95	2747.7	2987.3	6.1301
400	0.047 39	2892.9	3177.2	6.5408	0.039 93	2878.6	3158.1	6.4478	0.034 32	2863.8	3138.3	6.3634
450	0.052 14	2988.9	3301.8	6.7193	0.044 16	2978.0	3287.1	6.6327	0.038 17	2966.7	3272.0	6.5551
500	0.056 65	3082.2	3422.2	6.8803	0.048 14	3073.4	3410.3	6.7975	0.041 75	3064.3	3398.3	6.7240
550	0.061 01	3174.6	3540.6	7.0288	0.051 95	3167.2	3530.9	6.9486	0.045 16	3159.8	3521.0	6.8778
600	0.065 25	3266.9	3658.4	7.1677	0.055 65	3260.7	3650.3	7.0894	0.048 45	3254.4	3642.0	7.0206
700	0.073 52	3453.1	3894.2	7.4234	0.062 83	3448.5	3888.3	7.3476	0.054 81	3443.9	3882.4	7.2812
800	0.081 60	3643.1	4132.7	7.6566	0.069 81	3639.5	4128.2	7.5822	0.060 97	3636.0	4123.8	7.5173
900	0.089 58	3837.8	4375.3	7.8727	0.076 69	3835.0	4371.8	7.7991	0.067 02	3832.1	4368.3	7.7351
1000	0.097 49	4037.8	4622.7	8.0751	0.083 50	4035.3	4619.8	8.0020	0.073 01	4032.8	4616.9	7.9384
1100	0.105 36	4243.3	4875.4	8.2661	0.090 27	4240.9	4872.8	8.1933	0.078 96	4238.6	4870.3	8.1300
1200	0.113 21	4454.0	5133.3	8.4474	0.097 03	4451.7	5130.9	8.3747	0.084 89	4449.5	5128.5	8.3115
1300	0.121 06	4669.6	5396.0	8.6199	0.103 77	4667.3	5393.7	8.5475	0.090 80	4665.0	5391.5	8.4842
	P = 9.0 MPa (303.40°C)				*P* = 10.0 MPa (311.06°C)				*P* = 12.5 MPa (327.89°C)			
Sat.	0.020 48	2557.8	2742.1	5.6772	0.018 026	2544.4	2724.7	5.6141	0.013 495	2505.1	2673.8	5.4624
325	0.023 27	2646.6	2856.0	5.8712	0.019 861	2610.4	2809.1	5.7568				
350	0.025 80	2724.4	2956.6	6.0361	0.022 42	2699.2	2923.4	5.9443	0.016 126	2624.6	2826.2	5.7118
400	0.029 93	2848.4	3117.8	6.2854	0.026 41	2832.4	3096.5	6.2120	0.020 00	2789.3	3039.3	6.0417
450	0.033 50	2955.2	3256.6	6.4844	0.029 75	2943.4	3240.9	6.4190	0.022 99	2912.5	3199.8	6.2719
500	0.036 77	3055.2	3386.1	6.6576	0.032 79	3045.8	3373.7	6.5966	0.025 60	3021.7	3341.8	6.4618
550	0.039 87	3152.2	3511.0	6.8142	0.035 64	3144.6	3500.9	6.7561	0.028 01	3125.0	3475.2	6.6290
600	0.042 85	3248.1	3633.7	6.9589	0.038 37	3241.7	3625.3	6.9029	0.030 29	3225.4	3604.0	6.7810
650	0.045 74	3343.6	3755.3	7.0943	0.041 01	3338.2	3748.2	7.0398	0.032 48	3324.4	3730.4	6.9218
700	0.048 57	3439.3	3876.5	7.2221	0.043 58	3434.7	3870.5	7.1687	0.034 60	3422.9	3855.3	7.0536
800	0.054 09	3632.5	4119.3	7.4596	0.048 59	3628.9	4114.8	7.4077	0.038 69	3620.0	4103.6	7.2965
900	0.059 50	3829.2	4364.8	7.6783	0.053 49	3826.3	4361.2	7.6272	0.042 67	3819.1	4352.5	7.5182
1000	0.064 85	4030.3	4614.0	7.8821	0.058 32	4027.8	4611.0	7.8315	0.046 58	4021.6	4603.8	7.7237
1100	0.070 16	4236.3	4867.7	8.0740	0.063 12	4234.0	4865.1	8.0237	0.050 45	4228.2	4858.8	7.9165
1200	0.075 44	4447.2	5126.2	8.2556	0.067 89	4444.9	5123.8	8.2055	0.054 30	4439.3	5118.0	8.0937
1300	0.080 72	4662.7	5389.2	8.4284	0.072 65	4460.5	5387.0	8.3783	0.058 13	4654.8	5381.4	8.2717

T °C	v m³/kg	u kJ/kg	h kJ/kg	s kJ/(kg·K)	v m³/kg	u kJ/kg	h kJ/kg	s kJ/(kg·K)	v m³/kg	u kJ/kg	h kJ/kg	s kJ/(kg·K)
	P = 15.0 MPa (342.24°C)				**P = 17.5 MPa (354.75°C)**				**P = 20.0 MPa (365.81°C)**			
Sat.	0.010 337	2455.5	2610.5	5.3098	0.007 920	2390.2	2528.8	5.1419	0.005 834	2293.0	2409.7	4.9269
350	0.011 470	2520.4	2692.4	5.4421								
400	0.015 649	2740.7	2975.5	5.8811	0.012 447	2685.0	2902.9	5.7213	0.009 942	2619.3	2818.1	5.5540
450	0.018 445	2879.5	3156.2	6.1404	0.015 174	2844.2	3109.7	6.0184	0.012 695	2806.2	3060.1	5.9017
500	0.020 80	2996.6	3308.6	6.3443	0.017 358	2970.3	3274.1	6.2383	0.014 768	2942.9	3238.2	6.1401
550	0.022 93	3104.7	3448.6	6.5199	0.019 288	3083.9	3421.4	6.4230	0.016 555	3062.4	3393.5	6.3348
600	0.024 91	3208.6	3582.3	6.6776	0.021 06	3191.5	3560.1	6.5866	0.018 178	3174.0	3537.6	6.5048
650	0.026 80	3310.3	3712.3	6.8224	0.022 74	3296.0	3693.9	6.7357	0.019 693	3281.4	3675.3	6.6582
700	0.028 61	3410.9	3840.1	6.9572	0.024 34	3398.7	3824.6	6.8736	0.021 13	3386.4	3809.0	6.7993
800	0.032 10	3610.9	4092.4	7.2040	0.027 38	3601.8	4081.1	7.1244	0.023 85	3592.7	4069.7	7.0544
900	0.035 46	3811.9	4343.8	7.4279	0.030 31	3804.7	4335.1	7.3507	0.026 45	3797.5	4326.4	7.2830
1000	0.038 75	4015.4	4596.6	7.6348	0.033 16	4009.3	4589.5	7.5589	0.028 97	4003.1	4582.5	7.4925
1100	0.042 00	4222.6	4852.6	7.8283	0.035 97	4216.9	4846.4	7.7531	0.031 45	4211.3	4840.2	7.6874
1200	0.045 23	4433.8	5112.3	8.0108	0.038 76	4428.3	5106.6	7.9360	0.033 91	4422.8	5101.0	7.8707
1300	0.048 45	4649.1	5376.0	8.1840	0.041 54	4643.5	5370.5	8.1093	0.036 36	4638.0	5365.1	8.0442
	P = 25.0 MPa				**P = 30.0 MPa**				**P = 35.0 MPa**			
375	0.001 973 1	1798.7	1848.0	4.0320	0.001 789 2	1737.8	1791.5	3.9305	0.001 700 3	1702.9	1762.4	3.8722
400	0.006 004	2430.1	2580.2	5.1418	0.002 790	2067.4	2151.1	4.4728	0.002 100	1914.1	1987.6	4.2126
425	0.007 881	2609.2	2806.3	5.4723	0.005 303	2455.1	2614.2	5.1504	0.003 428	2253.4	2373.4	4.7747
450	0.009 162	2720.7	2949.7	5.6744	0.006 735	2619.3	2821.4	5.4424	0.004 961	2498.7	2672.4	5.1962
500	0.011 123	2884.3	3162.4	5.9592	0.008 678	2820.7	3081.1	5.7905	0.006 927	2751.9	2994.4	5.6282
550	0.012 724	3017.5	3335.6	6.1765	0.010 168	2970.3	3275.4	6.0342	0.008 345	2921.0	3213.0	5.9026
600	0.014 137	3137.9	3491.4	6.3602	0.011 446	3100.5	3443.9	6.2331	0.009 527	3062.0	3395.5	6.1179
650	0.015 433	3251.6	3637.4	6.5229	0.012 596	3221.0	3598.9	6.4058	0.010 575	3189.8	3559.9	6.3010
700	0.016 646	3361.3	3777.5	6.6707	0.013 661	3335.8	3745.6	6.5606	0.011 533	3309.8	3713.5	6.4631
800	0.018 912	3574.3	4047.1	6.9345	0.015 623	3555.5	4024.2	6.8332	0.013 278	3536.7	4001.5	6.7450
900	0.021 045	3783.0	4309.1	7.1680	0.017 448	3768.5	4291.9	7.0718	0.014 883	3754.0	4274.9	6.9386
1000	0.023 10	3990.9	4568.5	7.3802	0.019 196	3978.8	4554.7	7.2867	0.016 410	3966.7	4541.1	7.2064
1100	0.025 12	4200.2	4828.2	7.5765	0.020 903	4189.2	4816.3	7.4845	0.017 895	4178.3	4804.6	7.4037
1200	0.027 11	4412.0	5089.9	7.7605	0.022 589	4401.3	5079.0	7.6692	0.019 360	4390.7	5068.3	7.5910
1300	0.029 10	4626.9	5354.4	7.9342	0.024 266	4616.0	5344.0	7.8432	0.020 815	4605.1	5333.6	7.7653
	P = 40.0 MPa				**P = 50.0 MPa**				**P = 60.0 MPa**			
375	0.001 640 7	1677.1	1742.8	3.8290	0.001 559 4	1638.6	1716.6	3.7639	0.001 502 8	1609.4	1699.5	3.7141
400	0.001 907 7	1854.6	1930.9	4.1135	0.001 730 9	1788.1	1874.6	4.0031	0.001 633 5	1745.4	1843.4	3.9318
425	0.002 532	2096.9	2198.1	4.5029	0.002 007	1959.7	2060.0	4.2734	0.001 816 5	1892.7	2001.7	4.1626
450	0.003 693	2365.1	2512.8	4.9459	0.002 486	2159.6	2284.0	4.5884	0.002 085	2053.9	2179.0	4.4121
500	0.005 622	2678.4	2903.3	5.4700	0.003 892	2525.5	2720.1	5.1726	0.002 956	2390.6	2567.9	4.9321
550	0.006 984	2869.7	3149.1	5.7785	0.005 118	2763.6	3019.5	5.5485	0.003 956	2658.8	2896.2	5.3441
600	0.008 094	3022.6	3346.4	6.0144	0.006 112	2942.0	3247.6	5.8178	0.004 834	2861.1	3151.2	5.6452
650	0.009 063	3158.0	3520.6	6.2054	0.006 966	3093.5	3441.8	6.0342	0.005 595	3028.8	3364.5	5.8829
700	0.009 941	3283.6	3681.2	6.3750	0.007 727	3230.5	3616.8	6.2189	0.006 272	3177.2	3553.5	6.0824
800	0.011 523	3517.8	3978.7	6.6662	0.009 076	3479.8	3933.6	6.5290	0.007 459	3441.5	3889.1	6.4109
900	0.012 962	3739.4	4257.9	6.9150	0.010 283	3710.3	4224.4	6.7882	0.008 508	3681.0	4191.5	6.6805
1000	0.014 324	3954.6	4527.6	7.1356	0.011 411	3930.5	4501.1	7.0146	0.009 480	3906.4	4475.2	6.9127
1100	0.015 642	4167.4	4793.1	7.3364	0.012 496	4145.7	4770.5	7.2184	0.010 409	4124.1	4748.6	7.1195
1200	0.016 940	4380.1	5057.7	7.5224	0.013 561	4359.1	5037.2	7.4058	0.011 317	4338.2	5017.2	7.3083
1300	0.018 229	4594.3	5323.5	7.6969	0.014 616	4572.8	5303.6	7.5808	0.012 215	4551.4	5284.3	7.4837

TABLE A-7
Compressed liquid water

T °C	v m³/kg	u kJ/kg	h kJ/kg	s kJ/(kg·K)	v m³/kg	u kJ/kg	h kJ/kg	s kJ/(kg·K)	v m³/kg	u kJ/kg	h kJ/kg	s kJ/(kg·K)
	P = 5 MPa (263.99)°C				P = 10 MPa (311.06°C)				P = 15 MPa (342.24°C)			
Sat.	0.001 285 9	1147.8	1154.2	2.9202	0.001 452 4	1393.0	1407.6	3.3596	0.001 658 1	1585.6	1610.5	3.6848
0	0.000 997 7	0.04	5.04	0.0001	0.000 995 2	0.09	10.04	0.0002	0.000 992 8	0.15	15.05	0.0004
20	0.000 999 5	83.65	88.65	0.2956	0.000 997 2	83.36	93.33	0.2945	0.000 995 0	83.06	97.99	0.2934
40	0.001 005 6	166.95	171.97	0.5705	0.001 003 4	166.35	176.38	0.5686	0.001 001 3	165.76	180.78	0.5666
60	0.001 014 9	250.23	255.30	0.8285	0.001 012 7	249.36	259.49	0.8258	0.001 010 5	248.51	263.67	0.8232
80	0.001 026 8	333.72	338.85	1.0720	0.001 024 5	332.59	342.83	1.0688	0.001 022 2	331.48	346.81	1.0656
100	0.001 041 0	417.52	422.72	1.3030	0.001 038 5	416.12	426.50	1.2992	0.001 036 1	414.74	430.28	1.2955
120	0.001 057 6	501.80	507.09	1.5233	0.001 054 9	500.08	510.64	1.5189	0.001 052 2	498.40	514.19	1.5145
140	0.001 076 8	586.76	592.15	1.7343	0.001 073 7	584.68	595.42	1.7292	0.001 070 7	582.66	598.72	1.7242
160	0.001 098 8	672.62	678.12	1.9375	0.001 095 3	670.13	681.08	1.9317	0.001 091 8	667.71	684.09	1.9260
180	0.001 124 0	759.63	765.25	2.1341	0.001 119 9	756.65	767.84	2.1275	0.001 115 9	753.76	770.50	2.1210
200	0.001 153 0	848.1	853.9	2.3255	0.001 148 0	844.5	856.0	2.3178	0.001 143 3	841.0	858.2	2.3104
220	0.001 186 6	938.4	944.4	2.5128	0.001 180 5	934.1	945.9	2.5039	0.001 174 8	929.9	947.5	2.4953
240	0.001 226 4	1031.4	1037.5	2.6979	0.001 218 7	1026.0	1038.1	2.6872	0.001 211 4	1020.8	1039.0	2.6771
260	0.001 274 9	1127.9	1134.3	2.8830	0.001 264 5	1121.1	1133.7	2.8699	0.001 255 0	1114.6	1133.4	2.8576
280					0.001 321 6	1220.9	1234.1	3.0548	0.001 308 4	1212.5	1232.1	3.0393
300					0.001 397 2	1328.4	1342.3	3.2469	0.001 377 0	1316.6	1337.3	3.2260
320									0.001 472 4	1431.1	1453.2	3.4247
340									0.001 631 1	1567.5	1591.9	3.6546
	P = 20 MPa (365.81°C)				P = 30 MPa				P = 50 MPa			
Sat.	0.002 036	1785.6	1826.3	4.0139								
0	0.000 990 4	0.19	20.01	0.0004	0.000 985 6	0.25	29.82	0.0001	0.000 976 6	0.20	49.03	0.0014
20	0.000 992 8	82.77	102.62	0.2923	0.000 988 6	82.17	111.84	0.2899	0.000 980 4	81.00	130.02	0.2848
40	0.000 999 2	165.17	185.16	0.5646	0.000 995 1	164.04	193.89	0.5607	0.000 987 2	161.86	211.21	0.5527
60	0.001 008 4	247.68	267.85	0.8206	0.001 004 2	246.06	276.19	0.8154	0.000 996 2	242.98	292.79	0.8052
80	0.001 019 9	330.40	350.80	1.0624	0.001 015 6	328.30	358.77	1.0561	0.001 007 3	324.34	374.70	1.0440
100	0.001 033 7	413.39	434.06	1.2917	0.001 029 0	410.78	441.66	1.2844	0.001 020 1	405.88	456.89	1.2703
120	0.001 049 6	496.76	517.76	1.5102	0.001 044 5	493.59	524.93	1.5018	0.001 034 8	487.65	539.39	1.4857
140	0.001 067 8	580.69	602.04	1.7193	0.001 062 1	576.88	608.75	1.7098	0.001 051 5	569.77	622.35	1.6915
160	0.001 088 5	665.35	687.12	1.9204	0.001 082 1	660.82	693.28	1.9096	0.001 070 3	652.41	705.92	1.8891
180	0.001 112 0	750.95	773.20	2.1147	0.001 104 7	745.59	778.73	2.1024	0.001 091 2	735.69	790.25	2.0794
200	0.001 138 8	837.7	860.5	2.3031	0.001 130 2	831.4	865.3	2.2893	0.001 114 6	819.7	875.5	2.2634
220	0.001 169 5	925.9	949.3	2.4870	0.001 159 0	918.3	953.1	2.4711	0.001 140 8	904.7	961.7	2.4419
240	0.001 204 6	1016.0	1040.0	2.6674	0.001 192 0	1006.9	1042.6	2.6490	0.001 170 2	990.7	1049.2	2.6158
260	0.001 246 2	1108.6	1133.5	2.8459	0.001 230 3	1097.4	1134.3	2.8243	0.001 203 4	1078.1	1138.2	2.7860
280	0.001 296 5	1204.7	1230.6	3.0248	0.001 275 5	1190.7	1229.0	2.9986	0.001 241 5	1167.2	1229.3	2.9537
300	0.001 359 6	1306.1	1333.3	3.2071	0.001 330 4	1287.9	1327.8	3.1741	0.001 286 0	1258.7	1323.0	3.1200
320	0.001 443 7	1415.7	1444.6	3.3979	0.001 399 7	1390.7	1432.7	3.3539	0.001 338 8	1353.3	1420.2	3.2868
340	0.001 568 4	1539.7	1571.0	3.6075	0.001 492 0	1501.7	1546.5	3.5426	0.001 403 2	1452.0	1522.1	3.4557
360	0.001 822 6	1702.8	1739.3	3.8772	0.001 626 5	1626.6	1675.4	3.7494	0.001 483 8	1556.0	1630.2	3.6291
380					0.001 869 1	1781.4	1837.5	4.0012	0.001 588 4	1667.2	1746.6	3.8101

Temp. T °C	Sat. press. P_{sat} kPa	Specific volume m³/kg		Internal energy kJ/kg			Enthalpy kJ/kg			Entropy kJ/(kg · K)		
		Sat. ice $v_i \times 10^3$	Sat. vapor v_g	Sat. ice u_i	Subl. u_{ig}	Sat. vapor u_g	Sat. ice h_i	Subl. h_{ig}	Sat. vapor h_g	Sat. ice s_i	Subl. s_{ig}	Sat. vapor s_g
0.01	0.6113	1.0908	206.1	−333.40	2708.7	2375.3	−333.40	2834.8	2501.4	−1.221	10.378	9.156
0	0.6108	1.0908	206.3	−333.43	2708.8	2375.3	−333.43	2834.8	2501.3	−1.221	10.378	9.157
−2	0.5176	1.0904	241.7	−337.62	2710.2	2372.6	−337.62	2835.3	2497.7	−1.237	10.456	9.219
−4	0.4375	1.0901	283.8	−341.78	2711.6	2369.8	−341.78	2835.7	2494.0	−1.253	10.536	9.283
−6	0.3689	1.0898	334.2	−345.91	2712.9	2367.0	−345.91	2836.2	2490.3	−1.268	10.616	9.348
−8	0.3102	1.0894	394.4	−350.02	2714.2	2364.2	−350.02	2836.6	2486.6	−1.284	10.698	9.414
−10	0.2602	1.0891	466.7	−354.09	2715.5	2361.4	−354.09	2837.0	2482.9	−1.299	10.781	9.481
−12	0.2176	1.0888	553.7	−358.14	2716.8	2358.7	−358.14	2837.3	2479.2	−1.315	10.865	9.550
−14	0.1815	1.0884	658.8	−362.15	2718.0	2355.9	−362.15	2837.6	2475.5	−1.331	10.950	9.619
−16	0.1510	1.0881	786.0	−366.14	2719.2	2353.1	−366.14	2837.9	2471.8	−1.346	11.036	9.690
−18	0.1252	1.0878	940.5	−370.10	2720.4	2350.3	−370.10	2838.2	2468.1	−1.362	11.123	9.762
−20	0.1035	1.0874	1128.6	−374.03	2721.6	2347.5	−374.03	2838.4	2464.3	−1.377	11.212	9.835
−22	0.0853	1.0871	1358.4	−377.93	2722.7	2344.7	−377.93	2838.6	2460.6	−1.393	11.302	9.909
−24	0.0701	1.0868	1640.1	−381.80	2723.7	2342.0	−381.80	2838.7	2456.9	−1.408	11.394	9.985
−26	0.0574	1.0864	1986.4	−385.64	2724.8	2339.2	−385.64	2838.9	2453.2	−1.424	11.486	10.062
−28	0.0469	1.0861	2413.7	−389.45	2725.8	2336.4	−389.45	2839.0	2449.5	−1.439	11.580	10.141
−30	0.0381	1.0858	2943	−393.23	2726.8	2333.6	−393.23	2839.0	2445.8	−1.455	11.676	10.221
−32	0.0309	1.0854	3600	−396.98	2727.8	2330.8	−396.98	2839.1	2442.1	−1.471	11.773	10.303
−34	0.0250	1.0851	4419	−400.71	2728.7	2328.0	−400.71	2839.1	2438.4	−1.486	11.872	10.386
−36	0.0201	1.0848	5444	−404.40	2729.6	2325.2	−404.40	2839.1	2434.7	−1.501	11.972	10.470
−38	0.0161	1.0844	6731	−408.06	2730.5	2322.4	−408.06	2839.0	2430.9	−1.517	12.073	10.556
−40	0.0129	1.0841	8354	−411.70	2731.3	2319.6	−411.70	2839.9	2427.2	−1.532	12.176	10.644

T-s diagram for water. (*Source*: Lester Haar, John S. Gallagher, and George S. Kell, *NBS/NRC Steam Tables,* 1984. With permission from Hemisphere Publishing Corporation, New York.)

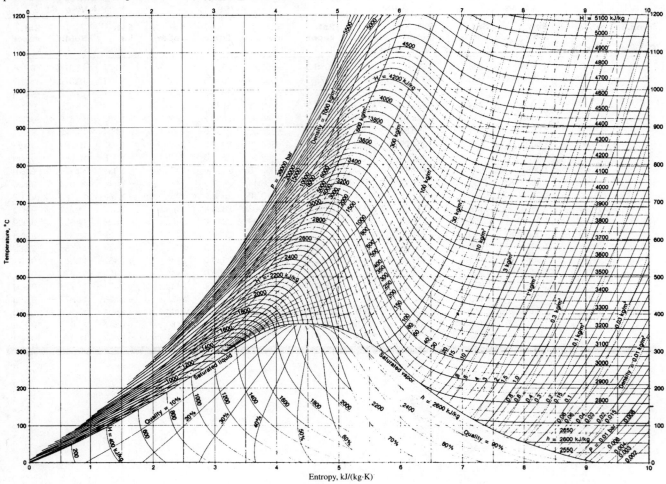

Mollier diagram for water. (*Source*: Lester Haar, John S. Gallagher, and George S. Kell, *NBS/NRC Steam Tables,* 1984. With permission from Hemisphere Publishing Corporation, New York.)

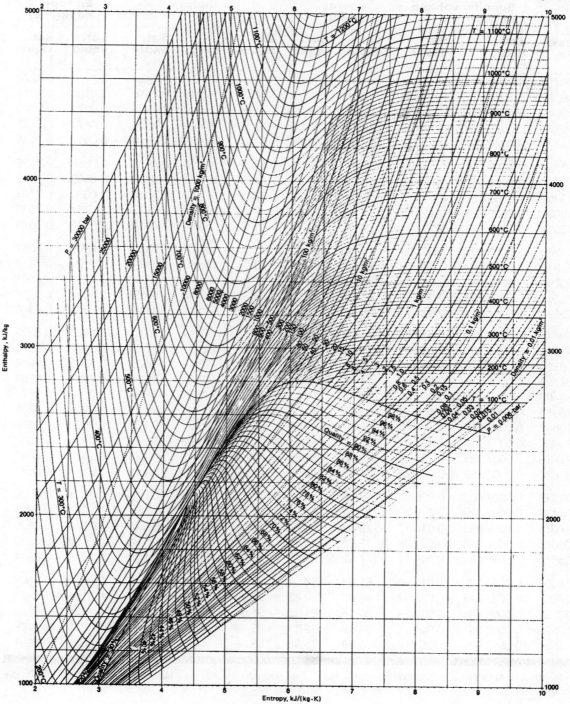

TABLE A-11
Saturated refrigerant-12–Temperature table

Temp. °C T	Sat. press. MPa P_{sat}	Specific volume m³/kg		Internal energy kJ/kg		Enthalpy kJ/kg			Entropy kJ/(kg·K)	
		Sat. liquid v_f	Sat. vapor v_g	Sat. liquid u_f	Sat. vapor u_g	Sat. liquid h_f	Evap. h_{fg}	Sat. vapor h_g	Sat. liquid s_f	Sat. vapor s_g
−40	0.06417	0.0006 595	0.241 91	−0.04	154.07	0	169.59	169.59	0	0.7274
−35	0.08071	0.0006 656	0.195 40	4.37	156.13	4.42	167.48	171.90	0.0187	0.7219
−30	0.10041	0.0006 720	0.159 38	8.79	158.20	8.86	165.33	174.20	0.0371	0.7170
−28	0.10927	0.0006 746	0.147 28	10.58	159.02	10.65	164.46	175.11	0.0444	0.7153
−26	0.11872	0.0006 773	0.136 28	12.35	159.84	12.43	163.59	176.02	0.0517	0.7135
−25	0.12368	0.0006 786	0.131 17	13.25	160.26	13.33	163.15	176.48	0.0552	0.7126
−24	0.12880	0.0006 800	0.126 28	14.13	160.67	14.22	162.71	176.93	0.0589	0.7119
−22	0.13953	0.0006 827	0.117 17	15.92	161.48	16.02	161.82	177.83	0.0660	0.7103
−20	0.15093	0.0006 855	0.108 85	17.72	162.31	17.82	160.92	178.74	0.0731	0.7087
−18	0.16304	0.0006 883	0.101 24	19.51	163.12	19.62	160.01	179.63	0.0802	0.7073
−15	0.18260	0.0006 926	0.091 02	22.20	164.35	22.33	158.64	180.97	0.0906	0.7051
−10	0.21912	0.0007 000	0.076 65	26.72	166.39	26.87	156.31	183.19	0.1080	0.7019
−5	0.26096	0.0007 078	0.064 96	31.27	168.42	31.45	153.93	185.37	0.1251	0.6991
0	0.30861	0.0007 159	0.055 39	35.83	170.44	36.05	151.48	187.53	0.1420	0.6965
4	0.35124	0.0007 227	0.048 95	39.51	172.04	39.76	149.47	189.23	0.1553	0.6946
8	0.39815	0.0007 297	0.043 40	43.21	173.63	43.50	147.41	190.91	0.1686	0.6929
12	0.44962	0.0007 370	0.038 60	46.93	175.20	47.26	145.30	192.56	0.1817	0.6913
16	0.50591	0.0007 446	0.034 42	50.67	176.78	51.05	143.14	194.19	0.1948	0.6898
20	0.56729	0.0007 525	0.030 78	54.44	178.32	54.87	140.91	195.78	0.2078	0.6884
24	0.63405	0.0007 607	0.027 59	58.25	179.85	58.73	138.61	197.34	0.2207	0.6871
26	0.66954	0.0007 650	0.026 14	60.17	180.61	60.68	137.44	198.11	0.2271	0.6865
28	0.70648	0.0007 694	0.024 78	62.09	181.36	62.63	136.24	198.87	0.2335	0.6859
30	0.74490	0.0007 739	0.023 51	64.01	182.11	64.59	135.03	199.62	0.2400	0.6853
32	0.78485	0.0007 785	0.022 31	65.96	182.85	66.57	133.79	200.36	0.2463	0.6847
34	0.82636	0.0007 832	0.021 18	67.90	183.59	68.55	132.53	201.09	0.2527	0.6842
36	0.86948	0.0007 880	0.020 12	69.86	184.31	70.55	131.25	201.80	0.2591	0.6836
38	0.91423	0.0007 929	0.019 12	71.84	185.03	72.56	129.94	202.51	0.2655	0.6831
40	0.96065	0.0007 980	0.018 17	73.82	185.74	74.59	128.61	203.20	0.2718	0.6825
42	1.0088	0.0008 033	0.017 28	75.82	186.45	76.63	127.25	203.88	0.2782	0.6820
44	1.0587	0.0008 086	0.016 44	77.82	187.13	78.68	125.87	204.54	0.2845	0.6814
48	1.1639	0.0008 199	0.014 88	81.88	188.51	82.83	123.00	205.83	0.2973	0.6802
52	1.2766	0.0008 318	0.013 49	86.00	189.83	87.06	119.99	207.05	0.3101	0.6791
56	1.3972	0.0008 445	0.012 24	90.18	191.10	91.36	116.84	208.20	0.3229	0.6779
60	1.5259	0.0008 581	0.011 11	94.43	192.31	95.74	113.52	209.26	0.3358	0.6765
112	4.1155	0.0017 92	0.001 79	175.98	175.98	183.35	0	183.35	0.5687	0.5687

Source: Tables A-11 through A-13 are adapted from Kenneth Wark, *Thermodynamics*, 4th ed., McGraw-Hill, New York, 1983, pp. 807–812. Originally published by E. I. du Pont de Nemours & Company, Inc., 1969.

Press. MPa P	Sat. temp. °C T_{sat}	Specific volume m³/kg		Internal energy kJ/kg		Enthalpy kJ/kg			Entropy kJ/(kg · K)	
		Sat. liquid v_f	Sat. vapor v_g	Sat. liquid u_f	Sat. vapor u_g	Sat. liquid h_f	Evap. h_{fg}	Sat. vapor h_g	Sat. liquid s_f	Sat. vapor s_g
0.06	−41.42	0.000 657 8	0.2575	−1.29	153.49	−1.25	170.19	168.94	−0.0054	0.7290
0.10	−30.10	0.000 671 9	0.1600	8.71	158.15	8.78	165.37	174.15	0.0368	0.7171
0.12	−25.74	0.000 677 6	0.1349	12.58	159.95	12.66	163.48	176.14	0.0526	0.7133
0.14	−21.91	0.000 682 8	0.1168	15.99	161.52	16.09	161.78	177.87	0.0663	0.7102
0.16	−18.49	0.000 687 6	0.1031	19.07	162.91	19.18	160.23	179.41	0.0784	0.7076
0.18	−15.38	0.000 692 1	0.092 25	21.86	164.19	21.98	158.82	180.80	0.0893	0.7054
0.20	−12.53	0.000 686 2	0.083 54	24.43	165.36	24.57	157.50	182.07	0.0992	0.7035
0.24	−7.42	0.000 704 0	0.070 33	29.06	167.44	29.23	155.09	184.32	0.1168	0.7004
0.28	−2.93	0.000 711 1	0.060 76	33.15	169.26	33.35	152.92	186.27	0.1321	0.6980
0.32	1.11	0.000 717 7	0.053 51	36.85	170.88	37.08	150.92	188.00	0.1457	0.6960
0.40	8.15	0.000 729 9	0.043 21	43.35	173.69	43.64	147.33	190.97	0.1691	0.6928
0.50	15.60	0.000 743 8	0.034 82	50.30	176.61	50.67	143.35	194.02	0.1935	0.6899
0.60	22.00	0.000 756 6	0.029 13	56.35	179.09	56.80	139.77	196.57	0.2142	0.6878
0.70	27.65	0.000 768 6	0.025 01	61.75	181.23	62.29	136.45	198.74	0.2324	0.6860
0.80	32.74	0.000 780 2	0.021 88	66.68	183.13	67.30	133.33	200.63	0.2487	0.6845
0.90	37.37	0.000 791 4	0.019 42	71.22	184.81	71.93	130.36	202.29	0.2634	0.6832
1.0	41.64	0.000 802 3	0.017 44	75.46	186.32	76.26	127.50	203.76	0.2770	0.6820
1.2	49.31	0.000 823 7	0.014 41	83.22	188.95	84.21	122.03	206.24	0.3015	0.6799
1.4	56.09	0.000 844 8	0.012 22	90.28	191.11	91.46	116.76	208.22	0.3232	0.6778
1.6	62.19	0.000 866 0	0.010 54	96.80	192.95	98.19	111.62	209.81	0.3329	0.6758

$$\Delta H = m(h_2 - h_1)$$

TABLE A-13
Superheated refrigerant-12

Temp. °C	v m³/kg	u kJ/kg	h kJ/kg	s kJ/(kg · K)	v m³/kg	u kJ/kg	h kJ/kg	s kJ/(kg · K)
	0.060 MPa (T_{sat} = −41.42°C)				**0.10 MPa** (T_{sat} = −30.10°C)			
Sat.	0.2575	153.49	168.94	0.7290	0.1600	158.15	174.15	0.7171
−40	0.2593	154.16	169.72	0.7324				
−20	0.2838	163.91	180.94	0.7785	0.1677	163.22	179.99	0.7406
0	0.3079	174.05	192.52	0.8225	0.1827	173.50	191.77	0.7854
10	0.3198	179.26	198.45	0.8439	0.1900	178.77	197.77	0.8070
20	0.3317	184.57	204.47	0.8647	0.1973	184.12	203.85	0.8281
30	0.3435	189.96	210.57	0.8852	0.2045	189.57	210.02	0.8488
40	0.3552	195.46	216.77	0.9053	0.2117	195.09	216.26	0.8691
50	0.3670	201.02	223.04	0.9251	0.2188	200.70	222.58	0.8889
60	0.3787	206.69	229.41	0.9444	0.2260	206.38	228.98	0.9084
80	0.4020	218.25	242.37	0.9822	0.2401	218.00	242.01	0.9464
	0.14 MPa (T_{sat} = −21.91°C)				**0.18 MPa** (T_{sat} = −15.38°C)			
Sat.	0.1168	161.52	177.87	0.7102	0.092.2	164.20	180.80	0.7054
−20	0.1179	162.50	179.01	0.7147				
−10	0.1235	167.69	184.97	0.7378	0.0925	164.39	181.03	0.7181
0	0.1289	172.94	190.99	0.7602	0.0991	172.37	190.21	0.7408
10	0.1343	178.28	197.08	0.7821	0.1034	177.77	196.38	0.7630
20	0.1397	183.67	203.23	0.8035	0.1076	183.23	202.60	0.7846
30	0.1449	189.17	209.46	0.8243	0.1118	188.77	208.89	0.8057
40	0.1502	194.72	215.75	0.8447	0.1160	194.35	215.23	0.8263
50	0.1553	200.38	222.12	0.8648	0.1201	200.02	221.64	0.8464
60	0.1605	206.08	228.55	0.8844	0.1241	205.78	228.12	0.8662
80	0.1707	217.74	241.64	0.9225	0.1322	217.47	241.27	0.9045
100	0.1809	229.67	255.00	0.9593	0.1402	229.45	254.69	0.9414
	0.20 MPa (T_{sat} = −12.53°C)				**0.24 MPa** (T_{sat} = −7.42°C)			
Sat.	0.0835	165.37	182.07	0.7035	0.0703	167.45	184.32	0.7004
0	0.0886	172.08	189.08	0.7325	0.0729	171.49	188.99	0.7177
10	0.0926	177.50	196.02	0.7548	0.0763	176.98	195.29	0.7404
20	0.0964	183.00	202.28	0.7766	0.0796	182.53	201.63	0.7624
30	0.1002	188.56	208.60	0.7978	0.0828	188.14	208.01	0.7838
40	0.1040	194.17	214.97	0.8184	0.0860	193.80	214.44	0.8047
50	0.1077	199.86	221.40	0.8387	0.0892	199.51	220.92	0.8251
60	0.1114	205.62	227.90	0.8585	0.0923	205.31	227.46	0.8450
80	0.1187	217.35	241.09	0.8969	0.0985	217.07	240.71	0.8836
100	0.1259	229.35	254.53	0.9339	0.1045	229.12	254.20	0.9208
120	0.1331	241.59	268.21	0.9696	0.1105	241.41	267.93	0.9566

Temp. °C	v m³/kg	u kJ/kg	h kJ/kg	s kJ/(kg·K)	v m³/kg	u kJ/kg	h kJ/kg	s kJ/(kg·K)
	0.28 MPa (T_{sat} = −2.93°C)				**0.32 MPa (T_{sat} = 1.11°C)**			
Sat.	0.060 76	169.26	186.27	0.6980	0.053 51	170.88	188.00	0.6960
0	0.061 66	170.89	188.15	0.7049				
10	0.064 64	176.45	194.55	0.7279	0.055 90	175.90	193.79	0.7167
20	0.067 55	182.06	200.97	0.7502	0.058 52	181.57	200.30	0.7393
30	0.070 40	187.71	207.42	0.7718	0.061 06	187.28	206.82	0.7612
40	0.073 19	193.42	213.91	0.7928	0.063 55	193.02	213.36	0.7824
50	0.075 94	199.18	220.44	0.8134	0.066 00	198.82	219.94	0.8031
60	0.078 65	205.00	227.02	0.8334	0.068 41	204.68	226.57	0.8233
80	0.083 99	216.82	240.34	0.8722	0.073 14	216.55	239.96	0.8623
100	0.089 24	228.29	253.88	0.9095	0.077 78	228.66	253.55	0.8997
120	0.094 43	241.21	267.65	0.9455	0.082 36	241.00	267.36	0.9358
	0.40 MPa (T_{sat} = 8.15°C)				**0.50 MPa (T_{sat} = 15.60°C)**			
Sat.	0.043 21	173.69	190.97	0.6928	0.034 82	176.61	194.02	0.6899
10	0.043 63	174.76	192.21	0.6972				
20	0.045 84	180.57	198.91	0.7204	0.035 65	179.26	197.08	0.7004
30	0.047 97	186.39	205.58	0.7428	0.037 46	185.23	203.96	0.7235
40	0.050 05	192.23	212.25	0.7645	0.039 22	191.20	210.81	0.7457
50	0.052 07	198.11	218.94	0.7855	0.040 91	197.19	217.64	0.7672
60	0.054 06	204.03	225.65	0.8060	0.042 57	203.20	224.48	0.7881
80	0.057 91	216.03	239.19	0.8454	0.045 78	215.32	238.21	0.8281
100	0.061 73	228.20	252.89	0.8831	0.048 89	227.61	252.05	0.8662
120	0.065 46	240.61	266.79	0.9194	0.051 93	240.10	266.06	0.9028
140	0.069 13	253.23	280.88	0.9544	0.054 92	252.77	280.23	0.9379
	0.60 MPa (T_{sat} = 22.00°C)				**0.70 MPa (T_{sat} = 27.65°C)**			
Sat.	0.029 13	179.09	196.57	0.6878	0.025 01	181.23	198.74	0.6860
30	0.030 42	184.01	202.26	0.7068	0.025 35	182.72	200.46	0.6917
40	0.031 97	190.13	209.31	0.7297	0.026 76	189.00	207.73	0.7153
50	0.033 45	196.23	216.30	0.7516	0.028 10	195.23	214.90	0.7378
60	0.034 89	202.34	223.27	0.7729	0.029 39	201.45	222.02	0.7595
80	0.037 65	214.61	237.20	0.8135	0.031 84	213.88	236.17	0.8008
100	0.040 32	227.01	251.20	0.8520	0.034 19	226.40	250.33	0.8398
120	0.042 91	239.57	265.32	0.8889	0.036 46	239.05	264.57	0.8769
140	0.045 45	252.31	279.58	0.9243	0.038 67	251.85	278.92	0.9125
160	0.047 94	265.25	294.01	0.9584	0.040 85	264.83	293.42	0.9468

TABLE A-13

(Continued)

Temp. °C	v m³/kg	u kJ/kg	h kJ/kg	s kJ/(kg · K)	v m³/kg	u kJ/kg	h kJ/kg	s kJ/(kg · K)
	0.80 MPa (T_{sat} = 32.74°C)				**0.90 MPa** (T_{sat} = 37.37°C)			
Sat.	0.021 88	183.13	200.63	0.6845	0.019 42	184.81	202.29	0.6832
40	0.022 83	187.81	206.07	0.7021	0.019 74	186.55	204.32	0.6897
50	0.024 07	194.19	213.45	0.7253	0.020 91	193.10	211.92	0.7136
60	0.025 25	200.52	220.72	0.7474	0.022 01	199.56	219.37	0.7363
80	0.027 48	213.13	235.11	0.7894	0.024 07	212.37	234.03	0.7790
100	0.029 59	225.77	249.44	0.8289	0.026 01	225.13	248.54	0.8190
120	0.031 62	238.51	263.81	0.8664	0.027 85	237.97	263.03	0.8569
140	0.033 59	251.39	278.26	0.9022	0.029 64	250.90	277.58	0.8930
160	0.035 52	264.41	292.83	0.9367	0.031 38	263.99	292.23	0.9276
180	0.037 42	277.60	307.54	0.9699	0.033 09	277.23	307.01	0.9609
	1.0 MPa (T_{sat} = 41.64°C)				**1.2 MPa** (T_{sat} = 49.31°C)			
Sat.	0.017 44	186.32	203.76	0.6820	0.014 41	188.95	206.24	0.6799
50	0.018 37	191.95	210.32	0.7026	0.014 48	189.43	206.81	0.6816
60	0.019 41	198.56	217.97	0.7259	0.015 46	196.41	214.96	0.7065
80	0.021 34	211.57	232.91	0.7695	0.017 22	209.91	230.57	0.7520
100	0.023 13	224.48	247.61	0.8100	0.018 81	223.13	245.70	0.7937
120	0.024 84	237.41	262.25	0.8482	0.020 30	236.27	260.53	0.8326
140	0.026 47	250.43	276.90	0.8845	0.021 72	249.45	275.51	0.8696
160	0.028 07	263.56	291.63	0.9193	0.023 09	263.70	290.41	0.9048
180	0.029 63	276.84	306.47	0.9528	0.024 43	276.05	305.37	0.9385
200	0.031 16	290.26	321.42	0.9851	0.025 74	289.55	320.44	0.9711
	1.4 MPa (T_{sat} = 56.09°C)				**1.6 MPa** (T_{sat} = 62.19°C)			
Sat.	0.012 22	191.11	208.22	0.6778	0.010 54	192.95	209.81	0.6758
60	0.012 58	194.00	211.61	0.6881				
80	0.014 25	208.11	228.06	0.7360	0.011 98	206.17	225.34	0.7209
100	0.015 71	221.70	243.69	0.7791	0.013 37	220.19	241.58	0.7656
120	0.017 05	235.09	258.96	0.8189	0.014 61	233.84	257.22	0.8065
140	0.018 32	248.43	274.08	0.8564	0.015 77	247.38	272.61	0.8447
160	0.019 54	261.80	289.16	0.8921	0.016 86	260.90	287.88	0.8808
180	0.020 71	275.27	304.26	0.9262	0.017 92	274.47	303.14	0.9152
200	0.021 86	288.84	319.44	0.9589	0.018 95	288.11	318.43	0.9482
220	0.022 99	302.51	334.70	0.9905	0.019 96	301.84	333.78	0.9800

P-h diagram for refrigerant-12. (Freon 12 is the du Pont trademark for refrigerant-12. Copyright E. I. du Pont de Nemours & Company; used with permission.)

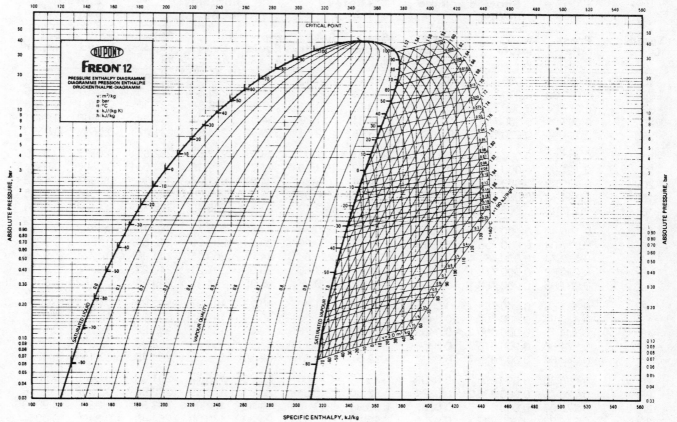

TABLE A-15a
Saturated refrigerant-134a–Temperature table

Temp. °C T	Press. MPa P_{sat}	Specific volume m³/kg Sat. liquid v_f	Sat. vapor v_g	Internal energy kJ/kg Sat. liquid u_f	Sat. vapor u_g	Enthalpy kJ/kg Sat. liquid h_f	Evap. h_{fg}	Sat. vapor h_g	Entropy kJ/(kg · K) Sat. liquid s_f	Sat. vapor s_g
−40	0.051 64	0.000 705 5	0.3569	−0.04	204.45	0.00	222.88	222.88	0.0000	0.9560
−36	0.063 32	0.000 711 3	0.2947	4.68	206.73	4.73	220.67	225.40	0.0201	0.9506
−32	0.077 04	0.000 717 2	0.2451	9.47	209.01	9.52	218.37	227.90	0.0401	0.9456
−28	0.093 05	0.000 723 3	0.2052	14.31	211.29	14.37	216.01	230.38	0.0600	0.9411
−26	0.101 99	0.000 726 5	0.1882	16.75	212.43	16.82	214.80	231.62	0.0699	0.9390
−24	0.111 60	0.000 729 6	0.1728	19.21	213.57	19.29	213.57	232.85	0.0798	0.9370
−22	0.121 92	0.000 732 8	0.1590	21.68	214.70	21.77	212.32	234.08	0.0897	0.9351
−20	0.132 99	0.000 736 1	0.1464	24.17	215.84	24.26	211.05	235.31	0.0996	0.9332
−18	0.144 83	0.000 739 5	0.1350	26.67	216.97	26.77	209.76	236.53	0.1094	0.9315
−16	0.157 48	0.000 742 8	0.1247	29.18	218.10	29.30	208.45	237.74	0.1192	0.9298
−12	0.185 40	0.000 749 8	0.1068	34.25	220.36	34.39	205.77	240.15	0.1388	0.9267
−8	0.217 04	0.000 756 9	0.0919	39.38	222.60	39.54	203.00	242.54	0.1583	0.9239
−4	0.252 74	0.000 764 4	0.0794	44.56	224.84	44.75	200.15	244.90	0.1777	0.9213
0	0.292 82	0.000 772 1	0.0689	49.79	227.06	50.02	197.21	247.23	0.1970	0.9190
4	0.337 65	0.000 780 1	0.0600	55.08	229.27	55.35	194.19	249.53	0.2162	0.9169
8	0.387 56	0.000 788 4	0.0525	60.43	231.46	60.73	191.07	251.80	0.2354	0.9150
12	0.442 94	0.000 797 1	0.0460	65.83	233.63	66.18	187.85	254.03	0.2545	0.9132
16	0.504 16	0.000 806 2	0.0405	71.29	235.78	71.69	184.52	256.22	0.2735	0.9116
20	0.571 60	0.000 815 7	0.0358	76.80	237.91	77.26	181.09	258.36	0.2924	0.9102
24	0.645 66	0.000 825 7	0.0317	82.37	240.01	82.90	177.55	260.45	0.3113	0.9089
26	0.685 30	0.000 830 9	0.0298	85.18	241.05	85.75	175.73	261.48	0.3208	0.9082
28	0.726 75	0.000 836 2	0.0281	88.00	242.08	88.61	173.89	262.50	0.3302	0.9076
30	0.770 06	0.000 841 7	0.0265	90.84	243.10	91.49	172.00	263.50	0.3396	0.9070
32	0.815 28	0.000 847 3	0.0250	93.70	244.12	94.39	170.09	264.48	0.3490	0.9064
34	0.862 47	0.000 853 0	0.0236	96.58	245.12	97.31	168.14	265.45	0.3584	0.9058
36	0.911 68	0.000 859 0	0.0223	99.47	246.11	100.25	166.15	266.40	0.3678	0.9053
38	0.962 98	0.000 865 1	0.0210	102.38	247.09	103.21	164.12	267.33	0.3772	0.9047
40	1.016 4	0.000 871 4	0.0199	105.30	248.06	106.19	162.05	268.24	0.3866	0.9041
42	1.072 0	0.000 878 0	0.0188	108.25	249.02	109.19	159.94	269.14	0.3960	0.9035
44	1.129 9	0.000 884 7	0.0177	111.22	249.96	112.22	157.79	270.01	0.4054	0.9030
48	1.252 6	0.000 898 9	0.0159	117.22	251.79	118.35	153.33	271.68	0.4243	0.9017
52	1.385 1	0.000 914 2	0.0142	123.31	253.55	124.58	148.66	273.24	0.4432	0.9004
56	1.527 8	0.000 930 8	0.0127	129.51	255.23	130.93	143.75	274.68	0.4622	0.8990
60	1.681 3	0.000 948 8	0.0114	135.82	256.81	137.42	138.57	275.99	0.4814	0.8973
70	2.116 2	0.001 002 7	0.0086	152.22	260.15	154.34	124.08	278.43	0.5302	0.8918
80	2.632 4	0.001 076 6	0.0064	169.88	262.14	172.71	106.41	279.12	0.5814	0.8827
90	3.243 5	0.001 194 9	0.0046	189.82	261.34	193.69	82.63	276.32	0.6380	0.8655
100	3.974 2	0.001 544 3	0.0027	218.60	248.49	224.74	34.40	259.13	0.7196	0.8117

Source: Tables A-15 and A-16 are adapted from M. J. Moran and H. N. Shapiro, *Fundamentals of Engineering Thermodynamics,* 2d ed., Wiley, New York, 1992, pp. 710–715. Originally based on equations from D. P. Wilson and R. S. Basu, "Thermodynamic Properties of a New Stratospherically Safe Working Fluid—Refrigerant 134a," *ASHRAE Trans.,* Vol. 94, Pt. 2, 1988, pp. 2095–2118.

Press. MPa P	Temp. °C T_{sat}	Specific volume m³/kg		Internal energy kJ/kg		Enthalpy kJ/kg			Entropy kJ/(kg·K)	
		Sat. liquid v_f	Sat. vapor v_g	Sat. liquid u_f	Sat. vapor u_g	Sat. liquid h_f	Evap. h_{fg}	Sat. vapor h_g	Sat. liquid s_f	Sat. vapor s_g
0.06	−37.07	0.000 709 7	0.3100	3.41	206.12	3.46	221.27	224.72	0.0147	0.9520
0.08	−31.21	0.000 718 4	0.2366	10.41	209.46	10.47	217.92	228.39	0.0440	0.9447
0.10	−26.43	0.000 725 8	0.1917	16.22	212.18	16.29	215.06	231.35	0.0678	0.9395
0.12	−22.36	0.000 732 3	0.1614	21.23	214.50	21.32	212.54	233.86	0.0879	0.9354
0.14	−18.80	0.000 738 1	0.1395	25.66	216.52	25.77	210.27	236.04	0.1055	0.9322
0.16	−15.62	0.000 743 5	0.1229	29.66	218.32	29.78	208.18	237.97	0.1211	0.9295
0.18	−12.73	0.000 748 5	0.1098	33.31	219.94	33.45	206.26	239.71	0.1352	0.9273
0.20	−10.09	0.000 753 2	0.0993	36.69	221.43	36.84	204.46	241.30	0.1481	0.9253
0.24	−5.37	0.000 761 8	0.0834	42.77	224.07	42.95	201.14	244.09	0.1710	0.9222
0.28	−1.23	0.000 769 7	0.0719	48.18	226.38	48.39	198.13	246.52	0.1911	0.9197
0.32	2.48	0.000 777 0	0.0632	53.06	228.43	53.31	195.35	248.66	0.2089	0.9177
0.36	5.84	0.000 783 9	0.0564	57.54	230.28	57.82	192.76	250.58	0.2251	0.9160
0.4	8.93	0.000 790 4	0.0509	61.69	231.97	62.00	190.32	252.32	0.2399	0.9145
0.5	15.74	0.000 805 6	0.0409	70.93	235.64	71.33	184.74	256.07	0.2723	0.9117
0.6	21.58	0.000 819 6	0.0341	78.99	238.74	79.48	179.71	259.19	0.2999	0.9097
0.7	26.72	0.000 832 8	0.0292	86.19	241.42	86.78	175.07	261.85	0.3242	0.9080
0.8	31.33	0.000 845 4	0.0255	92.75	243.78	93.42	170.73	264.15	0.3459	0.9066
0.9	35.53	0.000 857 6	0.0226	98.79	245.88	99.56	166.62	266.18	0.3656	0.9054
1.0	39.39	0.000 869 5	0.0202	104.42	247.77	105.29	162.68	267.97	0.3838	0.9043
1.2	46.32	0.000 892 8	0.0166	114.69	251.03	115.76	155.23	270.99	0.4164	0.9023
1.4	52.43	0.000 915 9	0.0140	123.98	253.74	125.26	148.14	273.40	0.4453	0.9003
1.6	57.92	0.000 939 2	0.0121	132.52	256.00	134.02	141.31	275.33	0.4714	0.8982
1.8	62.91	0.000 963 1	0.0105	140.49	257.88	142.22	134.60	276.83	0.4954	0.8959
2.0	67.49	0.000 987 8	0.0093	148.02	259.41	149.99	127.95	277.94	0.5178	0.8934
2.5	77.59	0.001 056 2	0.0069	165.48	261.84	168.12	111.06	279.17	0.5687	0.8854
3.0	86.22	0.001 141 6	0.0053	181.88	262.16	185.30	92.71	278.01	0.6156	0.8735

TABLE A-16
Superheated refrigerant 134a

T °C	v m³/kg	u kJ/kg	h kJ/kg	s kJ/(kg·K)	v m³/kg	u kJ/kg	h kJ/kg	s kJ/(kg·K)
	P = **0.06 MPa** (T_{sat} = −37.07°C)				P = **0.10 MPa** (T_{sat} = −26.43°C)			
Sat.	0.31003	206.12	224.72	0.9520	0.19170	212.18	231.35	0.9395
−20	0.33536	217.86	237.98	1.0062	0.19770	216.77	236.54	0.9602
−10	0.34992	224.97	245.96	1.0371	0.20686	224.01	244.70	0.9918
0	0.36433	232.24	254.10	1.0675	0.21587	231.41	252.99	1.0227
10	0.37861	239.69	262.41	1.0973	0.22473	238.96	261.43	1.0531
20	0.39279	247.32	270.89	1.1267	0.23349	246.67	270.02	1.0829
30	0.40688	255.12	279.53	1.1557	0.24216	254.54	278.76	1.1122
40	0.42091	263.10	288.35	1.1844	0.25076	262.58	287.66	1.1411
50	0.43487	271.25	297.34	1.2126	0.25930	270.79	296.72	1.1696
60	0.44879	279.58	306.51	1.2405	0.26779	279.16	305.94	1.1977
70	0.46266	288.08	315.84	1.2681	0.27623	287.70	315.32	1.2254
80	0.47650	296.75	325.34	1.2954	0.28464	296.40	324.87	1.2528
90	0.49031	305.58	335.00	1.3224	0.29302	305.27	334.57	1.2799
	P = **0.14 MPa** (T_{sat} = −18.80°C)				P = **0.18 MPa** (T_{sat} = −12.73°C)			
Sat.	0.13945	216.52	236.04	0.9322	0.10983	219.94	239.71	0.9273
−10	0.14549	223.03	243.40	0.9606	0.11135	222.02	242.06	0.9362
0	0.15219	230.55	251.86	0.9922	0.11678	229.67	250.69	0.9684
10	0.15875	238.21	260.43	1.0230	0.12207	237.44	259.41	0.9998
20	0.16520	246.01	269.13	1.0532	0.12723	245.33	268.23	1.0304
30	0.17155	253.96	277.97	1.0828	0.13230	253.36	277.17	1.0604
40	0.17783	262.06	286.96	1.1120	0.13730	261.53	286.24	1.0898
50	0.18404	270.32	296.09	1.1407	0.14222	269.85	295.45	1.1187
60	0.19020	278.74	305.37	1.1690	0.14710	278.31	304.79	1.1472
70	0.19633	287.32	314.80	1.1969	0.15193	286.93	314.28	1.1753
80	0.20241	296.06	324.39	1.2244	0.15672	295.71	323.92	1.2030
90	0.20846	304.95	334.14	1.2516	0.16148	304.63	333.70	1.2303
100	0.21449	314.01	344.04	1.2785	0.16622	313.72	343.63	1.2573
	P = **0.20 MPa** (T_{sat} = −10.09°C)				P = **0.24 MPa** (T_{sat} = −5.37°C)			
Sat.	0.09933	221.43	241.30	0.9253	0.08343	224.07	244.09	0.9222
−10	0.09938	221.50	241.38	0.9256				
0	0.10438	229.23	250.10	0.9582	0.08574	228.31	248.89	0.9399
10	0.10922	237.05	258.89	0.9898	0.08993	236.26	257.84	0.9721
20	0.11394	244.99	267.78	1.0206	0.09399	244.30	266.85	1.0034
30	0.11856	253.06	276.77	1.0508	0.09794	252.45	275.95	1.0339
40	0.12311	261.26	285.88	1.0804	0.10181	260.72	285.16	1.0637
50	0.12758	269.61	295.12	1.1094	0.10562	269.12	294.47	1.0930
60	0.13201	278.10	304.50	1.1380	0.10937	277.67	303.91	1.1218
70	0.13639	286.74	314.02	1.1661	0.11307	286.35	313.49	1.1501
80	0.14073	295.53	323.68	1.1939	0.11674	295.18	323.19	1.1780
90	0.14504	304.47	333.48	1.2212	0.12037	304.15	333.04	1.2055
100	0.14932	313.57	343.43	1.2483	0.12398	313.27	343.03	1.2326

T °C	v m³/kg	u kJ/kg	h kJ/kg	s kJ/(kg·K)	v m³/kg	u kJ/kg	h kJ/kg	s kJ/(kg·K)

T °C	v m³/kg	u kJ/kg	h kJ/kg	s kJ/(kg·K)	v m³/kg	u kJ/kg	h kJ/kg	s kJ/(kg·K)
	P = **0.28 MPa** (T_{sat} = **−1.23°C**)				P = **0.32 MPa** (T_{sat} = **2.48°C**)			
Sat.	0.07193	226.38	246.52	0.9197	0.06322	228.43	248.66	0.9177
0	0.07240	227.37	247.64	0.9238				
10	0.07613	235.44	256.76	0.9566	0.06576	234.61	255.65	0.9427
20	0.07972	243.59	265.91	0.9883	0.06901	242.87	264.95	0.9749
30	0.08320	251.83	275.12	1.0192	0.07214	251.19	274.28	1.0062
40	0.08660	260.17	284.42	1.0494	0.07518	259.61	283.67	1.0367
50	0.08992	268.64	293.81	1.0789	0.07815	268.14	293.15	1.0665
60	0.09319	277.23	303.32	1.1079	0.08106	276.79	302.72	1.0957
70	0.09641	285.96	312.95	1.1364	0.08392	285.56	312.41	1.1243
80	0.09960	294.82	322.71	1.1644	0.08674	294.46	322.22	1.1525
90	0.10275	303.83	332.60	1.1920	0.08953	303.50	332.15	1.1802
100	0.10587	312.98	342.62	1.2193	0.09229	312.68	342.21	1.2076
110	0.10897	322.27	352.78	1.2461	0.09503	322.00	352.40	1.2345
120	0.11205	331.71	363.08	1.2727	0.09774	331.45	362.73	1.2611
	P = **0.40 MPa** (T_{sat} = **8.93°C**)				P = **0.50 MPa** (T_{sat} = **15.74°C**)			
Sat.	0.05089	231.97	252.32	0.9145	0.04086	235.64	256.07	0.9117
10	0.05119	232.87	253.35	0.9182				
20	0.05397	241.37	262.96	0.9515	0.04188	239.40	260.34	0.9264
30	0.05662	249.89	272.54	0.8937	0.04416	248.20	270.28	0.9597
40	0.05917	258.47	282.14	1.0148	0.04633	256.99	280.16	0.9918
50	0.06164	267.13	291.79	1.0452	0.04842	265.83	290.04	1.0229
60	0.06405	275.89	301.51	1.0748	0.05043	274.73	299.95	1.0531
70	0.06641	284.75	311.32	1.1038	0.05240	283.72	309.92	1.0825
80	0.06873	293.73	321.23	1.1322	0.05432	292.80	319.96	1.1114
90	0.07102	302.84	331.25	1.1602	0.05620	302.00	330.10	1.1397
100	0.07327	312.07	341.38	1.1878	0.05805	311.31	340.33	1.1675
110	0.07550	321.44	351.64	1.2149	0.05988	320.74	350.68	1.1949
120	0.07771	330.94	362.03	1.2417	0.06168	330.30	361.14	1.2218
130	0.07991	340.58	372.54	1.2681	0.06347	339.98	371.72	1.2484
140	0.08208	350.35	383.18	1.2941	0.06524	349.79	382.42	1.2746

T °C	v m³/kg	u kJ/kg	h kJ/kg	s kJ/(kg·K)	v m³/kg	u kJ/kg	h kJ/kg	s kJ/(kg·K)

T °C	v m³/kg	u kJ/kg	h kJ/kg	s kJ/(kg·K)	v m³/kg	u kJ/kg	h kJ/kg	s kJ/(kg·K)
	P = 0.60 MPa (T_{sat} = 21.58°C)				P = 0.70 MPa (T_{sat} = 26.72°C)			
Sat.	0.03408	238.74	259.19	0.9097	0.02918	241.42	261.85	0.9080
30	0.03581	246.41	267.89	0.9388	0.02979	244.51	265.37	0.9197
40	0.03774	255.45	278.09	0.9719	0.03157	253.83	275.93	0.9539
50	0.03958	264.48	288.23	1.0037	0.03324	263.08	286.35	0.9867
60	0.04134	273.54	298.35	1.0346	0.03482	272.31	296.69	1.0182
70	0.04304	282.66	308.48	1.0645	0.03634	281.57	307.01	1.0487
80	0.04469	291.86	318.67	1.0938	0.03781	290.88	317.35	1.0784
90	0.04631	301.14	328.93	1.1225	0.03924	300.27	327.74	1.1074
100	0.04790	310.53	339.27	1.1505	0.04064	309.74	338.19	1.1358
110	0.04946	320.03	349.70	1.1781	0.04201	319.31	348.71	1.1637
120	0.05099	329.64	360.24	1.2053	0.04335	328.98	359.33	1.1910
130	0.05251	339.38	370.88	1.2320	0.04468	338.76	370.04	1.2179
140	0.05402	349.23	381.64	1.2584	0.04599	348.66	380.86	1.2444
150	0.05550	359.21	392.52	1.2844	0.04729	358.68	391.79	1.2706
160	0.05698	369.32	403.51	1.3100	0.04857	368.82	402.82	1.2963
	P = 0.80 MPa (T_{sat} = 31.33°C)				P = 0.90 MPa (T_{sat} = 35.53°C)			
Sat.	0.02547	243.78	264.15	0.9066	0.02255	245.88	266.18	0.9054
40	0.02691	252.13	273.66	0.9374	0.02325	250.32	271.25	0.9217
50	0.02846	261.62	284.39	0.9711	0.02472	260.09	282.34	0.9566
60	0.02992	271.04	294.98	1.0034	0.02609	269.72	293.21	0.9897
70	0.03131	280.45	305.50	1.0345	0.02738	279.30	303.94	1.0214
80	0.03264	289.89	316.00	1.0647	0.02861	288.87	314.62	1.0521
90	0.03393	299.37	326.52	1.0940	0.02980	298.46	325.28	1.0819
100	0.03519	308.93	337.08	1.1227	0.03095	308.11	335.96	1.1109
110	0.03642	318.57	347.71	1.1508	0.03207	317.82	346.68	1.1392
120	0.03762	328.31	358.40	1.1784	0.03316	327.62	357.47	1.1670
130	0.03881	338.14	369.19	1.2055	0.03423	337.52	368.33	1.1943
140	0.03997	348.09	380.07	1.2321	0.03529	347.51	379.27	1.2211
150	0.04113	358.15	391.05	1.2584	0.03633	357.61	390.31	1.2475
160	0.04227	368.32	402.14	1.2843	0.03736	367.82	401.44	1.2735
170	0.04340	378.61	413.33	1.3098	0.03838	378.14	412.68	1.2992
180	0.04452	389.02	424.63	1.3351	0.03939	388.57	424.02	1.3245

T °C	v m³/kg	u kJ/kg	h kJ/kg	s kJ/(kg·K)	v m³/kg	u kJ/kg	h kJ/kg	s kJ/(kg·K)
	P = **1.00 MPa** (T_{sat} = **39.39°C**)				P = **1.20 MPa** (T_{sat} = **46.32°C**)			
Sat.	0.02020	247.77	267.97	0.9043	0.01663	251.03	270.99	0.9023
40	0.02029	248.39	268.68	0.9066				
50	0.02171	258.48	280.19	0.9428	0.01712	254.98	275.52	0.9164
60	0.02301	268.35	291.36	0.9768	0.01835	265.42	287.44	0.9527
70	0.02423	278.11	302.34	1.0093	0.01947	275.59	298.96	0.9868
80	0.02538	287.82	313.20	1.0405	0.02051	285.62	310.24	1.0192
90	0.02649	297.53	324.01	1.0707	0.02150	295.59	321.39	1.0503
100	0.02755	307.27	334.82	1.1000	0.02244	305.54	332.47	1.0804
110	0.02858	317.06	345.65	1.1286	0.02335	315.50	343.52	1.1096
120	0.02959	326.93	356.52	1.1567	0.02423	325.51	354.58	1.1381
130	0.03058	336.88	367.46	1.1841	0.02508	335.58	365.68	1.1660
140	0.03154	346.92	378.46	1.2111	0.02592	345.73	376.83	1.1933
150	0.03250	357.06	389.56	1.2376	0.02674	355.95	388.04	1.2201
160	0.03344	367.31	400.74	1.2638	0.02754	366.27	399.33	1.2465
170	0.03436	377.66	412.02	1.2895	0.02834	376.69	410.70	1.2724
180	0.03528	388.12	423.40	1.3149	0.02912	387.21	422.16	1.2980
	P = **1.40 MPa** (T_{sat} = **52.43°C**)				P = **1.60 MPa** (T_{sat} = **57.92°C**)			
Sat.	0.01405	253.74	273.40	0.9003	0.01208	256.00	275.33	0.8982
60	0.01495	262.17	283.10	0.9297	0.01233	258.48	278.20	0.9069
70	0.01603	272.87	295.31	0.9658	0.01340	269.89	291.33	0.9457
80	0.01701	283.29	307.10	0.9997	0.01435	280.78	303.74	0.9813
90	0.01792	293.55	318.63	1.0319	0.01521	291.39	315.72	1.0148
100	0.01878	303.73	330.02	1.0628	0.01601	301.84	327.46	1.0467
110	0.01960	313.88	341.32	1.0927	0.01677	312.20	339.04	1.0773
120	0.02039	324.05	352.59	1.1218	0.01750	322.53	350.53	1.1069
130	0.02115	334.25	363.86	1.1501	0.01820	332.87	361.99	1.1357
140	0.02189	344.50	375.15	1.1777	0.01887	343.24	373.44	1.1638
150	0.02262	354.82	386.49	1.2048	0.01953	353.66	384.91	1.1912
160	0.02333	365.22	397.89	1.2315	0.02017	364.15	396.43	1.2181
170	0.02403	375.71	409.36	1.2576	0.02080	374.71	407.99	1.2445
180	0.02472	386.29	420.90	1.2834	0.02142	385.35	419.62	1.2704
190	0.02541	396.96	432.53	1.3088	0.02203	396.08	431.33	1.2960
200	0.02608	407.73	444.24	1.3338	0.02263	406.90	443.11	1.3212

TABLE A-17
Ideal-gas properties of air

T K	h kJ/kg	P_r	u kJ/kg	v_r	$s°$ kJ/(kg·K)	T K	h kJ/kg	P_r	u kJ/kg	v_r	$s°$ kJ/(kg·K)
200	199.97	0.3363	142.56	1707.0	1.295 59	580	586.04	14.38	419.55	115.7	2.373 48
210	209.97	0.3987	149.69	1512.0	1.344 44	590	596.52	15.31	427.15	110.6	2.391 40
220	219.97	0.4690	156.82	1346.0	1.391 05	600	607.02	16.28	434.78	105.8	2.409 02
230	230.02	0.5477	164.00	1205.0	1.435 57	610	617.53	17.30	442.42	101.2	2.426 44
240	240.02	0.6355	171.13	1084.0	1.478 24	620	628.07	18.36	450.09	96.92	2.443 56
250	250.05	0.7329	178.28	979.0	1.519 17	630	638.63	19.84	457.78	92.84	2.460 48
260	260.09	0.8405	185.45	887.8	1.558 48	640	649.22	20.64	465.50	88.99	2.477 16
270	270.11	0.9590	192.60	808.0	1.596 34	650	659.84	21.86	473.25	85.34	2.493 64
280	280.13	1.0889	199.75	738.0	1.632 79	660	670.47	23.13	481.01	81.89	2.509 85
285	285.14	1.1584	203.33	706.1	1.650 55	670	681.14	24.46	488.81	78.61	2.525 89
290	290.16	1.2311	206.91	676.1	1.668 02	680	691.82	25.85	496.62	75.50	2.541 75
295	295.17	1.3068	210.49	647.9	1.685 15	690	702.52	27.29	504.45	72.56	2.557 31
300	300.19	1.3860	214.07	621.2	1.702 03	700	713.27	28.80	512.33	69.76	2.572 77
305	305.22	1.4686	217.67	596.0	1.718 65	710	724.04	30.38	520.23	67.07	2.588 10
310	310.24	1.5546	221.25	572.3	1.734 98	720	734.82	32.02	528.14	64.53	2.603 19
315	315.27	1.6442	224.85	549.8	1.751 06	730	745.62	33.72	536.07	62.13	2.618 03
320	320.29	1.7375	228.42	528.6	1.766 90	740	756.44	35.50	544.02	59.82	2.632 80
325	325.31	1.8345	232.02	508.4	1.782 49	750	767.29	37.35	551.99	57.63	2.647 37
330	330.34	1.9352	235.61	489.4	1.797 83	760	778.18	39.27	560.01	55.54	2.661 76
340	340.42	2.149	242.82	454.1	1.827 90	780	800.03	43.35	576.12	51.64	2.690 13
350	350.49	2.379	250.02	422.2	1.857 08	800	821.95	47.75	592.30	48.08	2.717 87
360	360.58	2.626	257.24	393.4	1.885 43	820	843.98	52.59	608.59	44.84	2.745 04
370	370.67	2.892	264.46	367.2	1.913 13	840	866.08	57.60	624.95	41.85	2.771 70
380	380.77	3.176	271.69	343.4	1.940 01	860	888.27	63.09	641.40	39.12	2.797 83
390	390.88	3.481	278.93	321.5	1.966 33	880	910.56	68.98	657.95	36.61	2.823 44
400	400.98	3.806	286.16	301.6	1.991 94	900	932.93	75.29	674.58	34.31	2.848 56
410	411.12	4.153	293.43	283.3	2.016 99	920	955.38	82.05	691.28	32.18	2.873 24
420	421.26	4.522	300.69	266.6	2.041 42	940	977.92	89.28	708.08	30.22	2.897 48
430	431.43	4.915	307.99	251.1	2.065 33	960	1000.55	97.00	725.02	28.40	2.921 28
440	441.61	5.332	315.30	236.8	2.088 70	980	1023.25	105.2	741.98	26.73	2.944 68
450	451.80	5.775	322.62	223.6	2.111 61	1000	1046.04	114.0	758.94	25.17	2.967 70
460	462.02	6.245	329.97	211.4	2.134 07	1020	1068.89	123.4	776.10	23.72	2.990 34
470	472.24	6.742	337.32	200.1	2.156 04	1040	1091.85	133.3	793.36	22.39	3.012 60
480	482.49	7.268	344.70	189.5	2.177 60	1060	1114.86	143.9	810.62	21.14	3.034 49
490	492.74	7.824	352.08	179.7	2.198 76	1080	1137.89	155.2	827.88	19.98	3.056 08
500	503.02	8.411	359.49	170.6	2.219 52	1100	1161.07	167.1	845.33	18.896	3.077 32
510	513.32	9.031	366.92	162.1	2.239 93	1120	1184.28	179.7	862.79	17.886	3.098 25
520	523.63	9.684	374.36	154.1	2.259 97	1140	1207.57	193.1	880.35	16.946	3.118 83
530	533.98	10.37	381.84	146.7	2.279 67	1160	1230.92	207.2	897.91	16.064	3.139 16
540	544.35	11.10	389.34	139.7	2.299 06	1180	1254.34	222.2	915.57	15.241	3.159 16
550	554.74	11.86	396.86	133.1	2.318 09	1200	1277.79	238.0	933.33	14.470	3.178 88
560	565.17	12.66	404.42	127.0	2.336 85	1220	1301.31	254.7	951.09	13.747	3.198 34
570	575.59	13.50	411.97	121.2	2.355 31	1240	1324.93	272.3	968.95	13.069	3.217 51

898

T K	h kJ/kg	P_r	u kJ/kg	v_r	$s°$ kJ/(kg·K)	T K	h kJ/kg	P_r	u kJ/kg	v_r	$s°$ kJ/(kg·K)
1260	1348.55	290.8	986.90	12.435	3.236 38	1600	1757.57	791.2	1298.30	5.804	3.523 64
1280	1372.24	310.4	1004.76	11.835	3.255 10	1620	1782.00	834.1	1316.96	5.574	3.538 79
1300	1395.97	330.9	1022.82	11.275	3.273 45	1640	1806.46	878.9	1335.72	5.355	3.553 81
1320	1419.76	352.5	1040.88	10.747	3.291 60	1660	1830.96	925.6	1354.48	5.147	3.568 67
1340	1443.60	375.3	1058.94	10.247	3.309 59	1680	1855.50	974.2	1373.24	4.949	3.583 35
1360	1467.49	399.1	1077.10	9.780	3.327 24	1700	1880.1	1025	1392.7	4.761	3.5979
1380	1491.44	424.2	1095.26	9.337	3.344 74	1750	1941.6	1161	1439.8	4.328	3.6336
1400	1515.42	450.5	1113.52	8.919	3.362 00	1800	2003.3	1310	1487.2	3.944	3.6684
1420	1539.44	478.0	1131.77	8.526	3.379 01	1850	2065.3	1475	1534.9	3.601	3.7023
1440	1563.51	506.9	1150.13	8.153	3.395 86	1900	2127.4	1655	1582.6	3.295	3.7354
1460	1587.63	537.1	1168.49	7.801	3.412 47	1950	2189.7	1852	1630.6	3.022	3.7677
1480	1611.79	568.8	1186.95	7.468	3.428 92	2000	2252.1	2068	1678.7	2.776	3.7994
1500	1635.97	601.9	1205.41	7.152	3.445 16	2050	2314.6	2303	1726.8	2.555	3.8303
1520	1660.23	636.5	1223.87	6.854	3.461 20	2100	2377.4	2559	1775.3	2.356	3.8605
1540	1684.51	672.8	1242.43	6.569	3.477 12	2150	2440.3	2837	1823.8	2.175	3.8901
1560	1708.82	710.5	1260.99	6.301	3.492 76	2200	2503.2	3138	1872.4	2.012	3.9191
1580	1733.17	750.0	1279.65	6.046	3.508 29	2250	2566.4	3464	1921.3	1.864	3.9474

Source: Kenneth Wark, *Thermodynamics,* 4th ed., McGraw-Hill, New York, 1983, pp. 785–786, table A-5M. Originally published in J. H. Keenan and J. Kaye, *Gas Tables,* Wiley, New York, 1948.

TABLE A-18
Ideal-gas properties of nitrogen, N_2

T K	$\bar{h}$ kJ/kmol	$\bar{u}$ kJ/kmol	$\bar{s}°$ kJ/(kmol · K)	T K	$\bar{h}$ kJ/kmol	$\bar{u}$ kJ/kmol	$\bar{s}°$ kJ/(kmol · K)
0	0	0	0	600	17,563	12,574	212.066
220	6,391	4,562	182.639	610	17,864	12,792	212.564
230	6,683	4,770	183.938	620	18,166	13,011	213.055
240	6,975	4,979	185.180	630	18,468	13,230	213.541
250	7,266	5,188	186.370	640	18,772	13,450	214.018
260	7,558	5,396	187.514	650	19,075	13,671	214.489
270	7,849	5,604	188.614	660	19,380	13,892	214.954
280	8,141	5,813	189.673	670	19,685	14,114	215.413
290	8,432	6,021	190.695	680	19,991	14,337	215.866
298	8,669	6,190	191.502	690	20,297	14,560	216.314
300	8,723	6,229	191.682	700	20,604	14,784	216.756
310	9,014	6,437	192.638	710	20,912	15,008	217.192
320	9,306	6,645	193.562	720	21,220	15,234	217.624
330	9,597	6,853	194.459	730	21,529	15,460	218.059
340	9,888	7,061	195.328	740	21,839	15,686	218.472
350	10,180	7,270	196.173	750	22,149	15,913	218.889
360	10,471	7,478	196.995	760	22,460	16,141	219.301
370	10,763	7,687	197.794	770	22,772	16,370	219.709
380	11,055	7,895	198.572	780	23,085	16,599	220.113
390	11,347	8,104	199.331	790	23,398	16,830	220.512
400	11,640	8,314	200.071	800	23,714	17,061	220.907
410	11,932	8,523	200.794	810	24,027	17,292	221.298
420	12,225	8,733	201.499	820	24,342	17,524	221.684
430	12,518	8,943	202.189	830	24,658	17,757	222.067
440	12,811	9,153	202.863	840	24,974	17,990	222.447
450	13,105	9,363	203.523	850	25,292	18,224	222.822
460	13,399	9,574	204.170	860	25,610	18,459	223.194
470	13,693	9,786	204.803	870	25,928	18,695	223.562
480	13,988	9,997	205.424	880	26,248	18,931	223.927
490	14,285	10,210	206.033	890	26,568	19,168	224.288
500	14,581	10,423	206.630	900	26,890	19,407	224.647
510	14,876	10,635	207.216	910	27,210	19,644	225.002
520	15,172	10,848	207.792	920	27,532	19,883	225.353
530	15,469	11,062	208.358	930	27,854	20,122	225.701
540	15,766	11,277	208.914	940	28,178	20,362	226.047
550	16,064	11,492	209.461	950	28,501	20,603	226.389
560	16,363	11,707	209.999	960	28,826	20,844	226.728
570	16,662	11,923	210.528	970	29,151	21,086	227.064
580	16,962	12,139	211.049	980	29,476	21,328	227.398
590	17,262	12,356	211.562	990	29,803	21,571	227.728

T K	$\bar{h}$ kJ/kmol	$\bar{u}$ kJ/kmol	$\bar{s}^\circ$ kJ/(kmol · K)	T K	$\bar{h}$ kJ/kmol	$\bar{u}$ kJ/kmol	$\bar{s}^\circ$ kJ/(kmol · K)
1000	30,129	21,815	228.057	1760	56,227	41,594	247.396
1020	30,784	22,304	228.706	1780	56,938	42,139	247.798
1040	31,442	22,795	229.344	1800	57,651	42,685	248.195
1060	32,101	23,288	229.973	1820	58,363	43,231	248.589
1080	32,762	23,782	230.591	1840	59,075	43,777	248.979
1100	33,426	24,280	231.199	1860	59,790	44,324	249.365
1120	34,092	24,780	231.799	1880	60,504	44,873	249.748
1140	34,760	25,282	232.391	1900	61,220	45,423	250.128
1160	35,430	25,786	232.973	1920	61,936	45,973	250.502
1180	36,104	26,291	233.549	1940	62,654	46,524	250.874
1200	36,777	26,799	234.115	1960	63,381	47,075	251.242
1220	37,452	27,308	234.673	1980	64,090	47,627	251.607
1240	38,129	27,819	235.223	2000	64,810	48,181	251.969
1260	38,807	28,331	235.766	2050	66,612	49,567	252.858
1280	39,488	28,845	236.302	2100	68,417	50,957	253.726
1300	40,170	29,361	236.831	2150	70,226	52,351	254.578
1320	40,853	29,378	237.353	2200	72,040	53,749	255.412
1340	41,539	30,398	237.867	2250	73,856	55,149	256.227
1360	42,227	30,919	238.376	2300	75,676	56,553	257.027
1380	42,915	31,441	238.878	2350	77,496	57,958	257.810
1400	43,605	31,964	239.375	2400	79,320	59,366	258.580
1420	44,295	32,489	239.865	2450	81,149	60,779	259.332
1440	44,988	33,014	240.350	2500	82,981	62,195	260.073
1460	45,682	33,543	240.827	2550	84,814	63,613	260.799
1480	46,377	34,071	241.301	2600	86,650	65,033	261.512
1500	47,073	34,601	241.768	2650	88,488	66,455	262.213
1520	47,771	35,133	242.228	2700	90,328	67,880	262.902
1540	48,470	35,665	242.685	2750	92,171	69,306	263.577
1560	49,168	36,197	243.137	2800	94,014	70,734	264.241
1580	49,869	36,732	243.585	2850	95,859	72,163	264.895
1600	50,571	37,268	244.028	2900	97,705	73,593	265.538
1620	51,275	37,806	244.464	2950	99,556	75,028	266.170
1640	51,980	38,344	244.896	3000	101,407	76,464	266.793
1660	52,686	38,884	245.324	3050	103,260	77,902	267.404
1680	53,393	39,424	245.747	3100	105,115	79,341	268.007
1700	54,099	39,965	246.166	3150	106,972	80,782	268.601
1720	54,807	40,507	246.580	3200	108,830	82,224	269.186
1740	55,516	41,049	246.990	3250	110,690	83,668	269.763

Source: Tables A-18 through A-25 are adapted from Kenneth Wark, *Thermodynamics*, 4th ed., McGraw-Hill, New York, 1983, pp. 787–798. Originally published in JANAF, *Thermochemical Tables*, NSRDS-NBS-37, 1971.

TABLE A-19
Ideal-gas properties of oxygen, O_2

T K	$\bar{h}$ kJ/kmol	$\bar{u}$ kJ/kmol	$\bar{s}^\circ$ kJ/(kmol · K)	T K	$\bar{h}$ kJ/kmol	$\bar{u}$ kJ/kmol	$\bar{s}^\circ$ kJ/(kmol · K)
0	0	0	0	600	17,929	12,940	226.346
220	6,404	4,575	196.171	610	18,250	13,178	226.877
230	6,694	4,782	197.461	620	18,572	13,417	227.400
240	6,984	4,989	198.696	630	18,895	13,657	227.918
250	7,275	5,197	199.885	640	19,219	13,898	228.429
260	7,566	5,405	201.027	650	19,544	14,140	228.932
270	7,858	5,613	202.128	660	19,870	14,383	229.430
280	8,150	5,822	203.191	670	20,197	14,626	229.920
290	8,443	6,032	204.218	680	20,524	14,871	230.405
298	8,682	6,203	205.033	690	20,854	15,116	230.885
300	8,736	6,242	205.213	700	21,184	15,364	231.358
310	9,030	6,453	206.177	710	21,514	15,611	231.827
320	9,325	6,664	207.112	720	21,845	15,859	232.291
330	9,620	6,877	208.020	730	22,177	16,107	232.748
340	9,916	7,090	208.904	740	22,510	16,357	233.201
350	10,213	7,303	209.765	750	22,844	16,607	233.649
360	10,511	7,518	210.604	760	23,178	16,859	234.091
370	10,809	7,733	211.423	770	23,513	17,111	234.528
380	11,109	7,949	212.222	780	23,850	17,364	234.960
390	11,409	8,166	213.002	790	24,186	17,618	235.387
400	11,711	8,384	213.765	800	24,523	17,872	235.810
410	12,012	8,603	214.510	810	24,861	18,126	236.230
420	12,314	8,822	215.241	820	25,199	18,382	236.644
430	12,618	9,043	215.955	830	25,537	18,637	237.055
440	12,923	9,264	216.656	840	25,877	18,893	237.462
450	13,228	9,487	217.342	850	26,218	19,150	237.864
460	13,525	9,710	218.016	860	26,559	19,408	238.264
470	13,842	9,935	218.676	870	26,899	19,666	238.660
480	14,151	10,160	219.326	880	27,242	19,925	239.051
490	14,460	10,386	219.963	890	27,584	20,185	239.439
500	14,770	10,614	220.589	900	27,928	20,445	239.823
510	15,082	10,842	221.206	910	28,272	20,706	240.203
520	15,395	11,071	221.812	920	28,616	20,967	240.580
530	15,708	11,301	222.409	930	28,960	21,228	240.953
540	16,022	11,533	222.997	940	29,306	21,491	241.323
550	16,338	11,765	223.576	950	29,652	21,754	241.689
560	16,654	11,998	224.146	960	29,999	22,017	242.052
570	16,971	12,232	224.708	970	30,345	22,280	242.411
580	17,290	12,467	225.262	980	30,692	22,544	242.768
590	17,609	12,703	225.808	990	31,041	22,809	243.120

T K	$\bar{h}$ kJ/kmol	$\bar{u}$ kJ/kmol	$\bar{s}°$ kJ/(kmol · K)	T K	$\bar{h}$ kJ/kmol	$\bar{u}$ kJ/kmol	$\bar{s}°$ kJ/(kmol · K)
1000	31,389	23,075	243.471	1760	58,880	44,247	263.861
1020	32,088	23,607	244.164	1780	59,624	44,825	264.283
1040	32,789	24,142	244.844	1800	60,371	45,405	264.701
1060	33,490	24,677	245.513	1820	61,118	45,986	265.113
1080	34,194	25,214	246.171	1840	61,866	46,568	265.521
1100	34,899	25,753	246.818	1860	62,616	47,151	265.925
1120	35,606	26,294	247.454	1880	63,365	47,734	266.326
1140	36,314	26,836	248.081	1900	64,116	48,319	266.722
1160	37,023	27,379	248.698	1920	64,868	48,904	267.115
1180	37,734	27,923	249.307	1940	65,620	49,490	267.505
1200	38,447	28,469	249.906	1960	66,374	50,078	267.891
1220	39,162	29,018	250.497	1980	67,127	50,665	268.275
1240	39,877	29,568	251.079	2000	67,881	51,253	268.655
1260	40,594	30,118	251.653	2050	69,772	52,727	269.588
1280	41,312	30,670	252.219	2100	71,668	54,208	270.504
1300	42,033	31,224	252.776	2150	73,573	55,697	271.399
1320	42,753	31,778	253.325	2200	75,484	57,192	272.278
1340	43,475	32,334	253.868	2250	77,397	58,690	273.136
1360	44,198	32,891	254.404	2300	79,316	60,193	273.891
1380	44,923	33,449	254.932	2350	81,243	61,704	274.809
1400	45,648	34,008	255.454	2400	83,174	63,219	275.625
1420	46,374	34,567	255.968	2450	85,112	64,742	276.424
1440	47,102	35,129	256.475	2500	87,057	66,271	277.207
1460	47,831	35,692	256.978	2550	89,004	67,802	277.979
1480	48,561	36,256	257.474	2600	90,956	69,339	278.738
1500	49,292	36,821	257.965	2650	92,916	70,883	279.485
1520	50,024	37,387	258.450	2700	94,881	72,433	280.219
1540	50,756	37,952	258.928	2750	96,852	73,987	280.942
1560	51,490	38,520	259.402	2800	98,826	75,546	281.654
1580	52,224	39,088	259.870	2850	100,808	77,112	282.357
1600	52,961	39,658	260.333	2900	102,793	78,682	283.048
1620	53,696	40,227	260.791	2950	104,785	80,258	283.728
1640	54,434	40,799	261.242	3000	106,780	81,837	284.399
1660	55,172	41,370	261.690	3050	108,778	83,419	285.060
1680	55,912	41,944	262.132	3100	110,784	85,009	285.713
1700	56,652	42,517	262.571	3150	112,795	86,601	286.355
1720	57,394	43,093	263.005	3200	114,809	88,203	286.989
1740	58,136	43,669	263.435	3250	116,827	89,804	287.614

TABLE A-20
Ideal-gas properties of carbon dioxide, CO_2

T K	$\bar{h}$ kJ/kmol	$\bar{u}$ kJ/kmol	$\bar{s}°$ kJ/(kmol · K)	T K	$\bar{h}$ kJ/kmol	$\bar{u}$ kJ/kmol	$\bar{s}°$ kJ/(kmol · K)
0	0	0	0	600	22,280	17,291	243.199
220	6,601	4,772	202.966	610	22,754	17,683	243.983
230	6,938	5,026	204.464	620	23,231	18,076	244.758
240	7,280	5,285	205.920	630	23,709	18,471	245.524
250	7,627	5,548	207.337	640	24,190	18,869	246.282
260	7,979	5,817	208.717	650	24,674	19,270	247.032
270	8,335	6,091	210.062	660	25,160	19,672	247.773
280	8,697	6,369	211.376	670	25,648	20,078	248.507
290	9,063	6,651	212.660	680	26,138	20,484	249.233
298	9,364	6,885	213.685	690	26,631	20,894	249.952
300	9,431	6,939	213.915	700	27,125	21,305	250.663
310	9,807	7,230	215.146	710	27,622	21,719	251.368
320	10,186	7,526	216.351	720	28,121	22,134	252.065
330	10,570	7,826	217.534	730	28,622	22,552	252.755
340	10,959	8,131	218.694	740	29,124	22,972	253.439
350	11,351	8,439	219.831	750	29,629	23,393	254.117
360	11,748	8,752	220.948	760	20,135	23,817	254.787
370	12,148	9,068	222.044	770	30,644	24,242	255.452
380	12,552	9,392	223.122	780	31,154	24,669	256.110
390	12,960	9,718	224.182	790	31,665	25,097	256.762
400	13,372	10,046	225.225	800	32,179	25,527	257.408
410	13,787	10,378	226.250	810	32,694	25,959	258.048
420	14,206	10,714	227.258	820	33,212	26,394	258.682
430	14,628	11,053	228.252	830	33,730	26,829	259.311
440	15,054	11,393	229.230	840	34,251	27,267	259.934
450	15,483	11,742	230.194	850	34,773	27,706	260.551
460	15,916	12,091	231.144	860	35,296	28,125	261.164
470	16,351	12,444	232.080	870	35,821	28,588	261.770
480	16,791	12,800	233.004	880	36,347	29,031	262.371
490	17,232	13,158	233.916	890	36,876	29,476	262.968
500	17,678	13,521	234.814	900	37,405	29,922	263.559
510	18,126	13,885	235.700	910	37,935	30,369	264.146
520	18,576	14,253	236.575	920	38,467	30,818	264.728
530	19,029	14,622	237.439	930	39,000	31,268	265.304
540	19,485	14,996	238.292	940	39,535	31,719	265.877
550	19,945	15,372	239.135	950	40,070	32,171	266.444
560	20,407	15,751	239.962	960	40,607	32,625	267.007
570	20,870	16,131	240.789	970	41,145	33,081	267.566
580	21,337	16,515	241.602	980	41,685	33,537	268.119
590	21,807	16,902	242.405	990	42,226	33,995	268.670

T K	$\bar{h}$ kJ/kmol	$\bar{u}$ kJ/kmol	$\bar{s}^\circ$ kJ/(kmol · K)	T K	$\bar{h}$ kJ/kmol	$\bar{u}$ kJ/kmol	$\bar{s}^\circ$ kJ/(kmol · K)
1000	42,769	34,455	269.215	1760	86,420	71,787	301.543
1020	43,859	35,378	270.293	1780	87,612	72,812	302.217
1040	44,953	36,306	271.354	1800	88,806	73,840	302.884
1060	46,051	37,238	272.400	1820	90,000	74,868	303.544
1080	47,153	38,174	273.430	1840	91,196	75,897	304.198
1100	48,258	39,112	274.445	1860	92,394	76,929	304.845
1120	49,369	40,057	275.444	1880	93,593	77,962	305.487
1140	50,484	41,006	276.430	1900	94,793	78,996	306.122
1160	51,602	41,957	277.403	1920	95,995	80,031	306.751
1180	52,724	42,913	278.361	1940	97,197	81,067	307.374
1200	53,848	43,871	279.307	1960	98,401	82,105	307.992
1220	54,977	44,834	280.238	1980	99,606	83,144	308.604
1240	56,108	45,799	281.158	2000	100,804	84,185	309.210
1260	57,244	46,768	282.066	2050	103,835	86,791	310.701
1280	58,381	47,739	282.962	2100	106,864	89,404	312.160
1300	59,522	48,713	283.847	2150	109,898	92,023	313.589
1320	60,666	49,691	284.722	2200	112,939	94,648	314.988
1340	61,813	50,672	285.586	2250	115,984	97,277	316.356
1360	62,963	51,656	286.439	2300	119,035	99,912	317.695
1380	64,116	52,643	287.283	2350	122,091	102,552	319.011
1400	65,271	53,631	288.106	2400	125,152	105,197	320.302
1420	66,427	54,621	288.934	2450	128,219	107,849	321.566
1440	67,586	55,614	289.743	2500	131,290	110,504	322.808
1460	68,748	56,609	290.542	2550	134,368	113,166	324.026
1480	66,911	57,606	291.333	2600	137,449	115,832	325.222
1500	71,078	58,606	292.114	2650	140,533	118,500	326.396
1520	72,246	59,609	292.888	2700	143,620	121,172	327.549
1540	73,417	60,613	292.654	2750	146,713	123,849	328.684
1560	74,590	61,620	294.411	2800	149,808	126,528	329.800
1580	76,767	62,630	295.161	2850	152,908	129,212	330.896
1600	76,944	63,741	295.901	2900	156,009	131,898	331.975
1620	78,123	64,653	296.632	2950	159,117	134,589	333.037
1640	79,303	65,668	297.356	3000	162,226	137,283	334.084
1660	80,486	66,592	298.072	3050	165,341	139,982	335.114
1680	81,670	67,702	298.781	3100	168,456	142,681	336.126
1700	82,856	68,721	299.482	3150	171,576	145,385	337.124
1720	84,043	69,742	300.177	3200	174,695	148,089	338.109
1740	85,231	70,764	300.863	3250	177,822	150,801	339.069

TABLE A-21
Ideal-gas properties of carbon monoxide, CO

T K	$\bar{h}$ kJ/kmol	$\bar{u}$ kJ/kmol	$\bar{s}°$ kJ/(kmol · K)	T K	$\bar{h}$ kJ/kmol	$\bar{u}$ kJ/kmol	$\bar{s}°$ kJ/(kmol · K)
0	0	0	0	600	17,611	12,622	218.204
220	6,391	4,562	188.683	610	17,915	12,843	218.708
230	6,683	4,771	189.980	620	18,221	13,066	219.205
240	6,975	4,979	191.221	630	18,527	13,289	219.695
250	7,266	5,188	192.411	640	18,833	13,512	220.179
260	7,558	5,396	193.554	650	19,141	13,736	220.656
270	7,849	5,604	194.654	660	19,449	13,962	221.127
280	8,140	5,812	195.713	670	19,758	14,187	221.592
290	8,432	6,020	196.735	680	20,068	14,414	222.052
298	8,669	6,190	197.543	690	20,378	14,641	222.505
300	8,723	6,229	197.723	700	20,690	14,870	222.953
310	9,014	6,437	198.678	710	21,002	15,099	223.396
320	9,306	6,645	199.603	720	21,315	15,328	223.833
330	9,597	6,854	200.500	730	21,628	15,558	224.265
340	9,889	7,062	201.371	740	21,943	15,789	224.692
350	10,181	7,271	202.217	750	22,258	16,022	225.115
360	10,473	7,480	203.040	760	22,573	16,255	225.533
370	10,765	7,689	203.842	770	22,890	16,488	225.947
380	11,058	7,899	204.622	780	23,208	16,723	226.357
390	11,351	8,108	205.383	790	23,526	16,957	226.762
400	11,644	8,319	206.125	800	23,844	17,193	227.162
410	11,938	8,529	206.850	810	24,164	17,429	227.559
420	12,232	8,740	207.549	820	24,483	17,665	227.952
430	12,526	8,951	208.252	830	24,803	17,902	228.339
440	12,821	9,163	208.929	840	25,124	18,140	228.724
450	13,116	9,375	209.593	850	25,446	18,379	229.106
460	13,412	9,587	210.243	860	25,768	18,617	229.482
470	13,708	9,800	210.880	870	26,091	18,858	229.856
480	14,005	10,014	211.504	880	26,415	19,099	230.227
490	14,302	10,228	212.117	890	26,740	19,341	230.593
500	14,600	10,443	212.719	900	27,066	19,583	230.957
510	14,898	10,658	213.310	910	27,392	19,826	231.317
520	15,197	10,874	213.890	920	27,719	20,070	231.674
530	15,497	11,090	214.460	930	28,046	20,314	232.028
540	15,797	11,307	215.020	940	28,375	20,559	232.379
550	16,097	11,524	215.572	950	28,703	20,805	232.727
560	16,399	11,743	216.115	960	29,033	21,051	233.072
570	16,701	11,961	216.649	970	29,362	21,298	233.413
580	17,003	12,181	217.175	980	29,693	21,545	233.752
590	17,307	12,401	217.693	990	30,024	21,793	234.088

T K	$\bar{h}$ kJ/kmol	$\bar{u}$ kJ/kmol	$\bar{s}°$ kJ/(kmol · K)	T K	$\bar{h}$ kJ/kmol	$\bar{u}$ kJ/kmol	$\bar{s}°$ kJ/(kmol · K)
1000	30,355	22,041	234.421	1760	56,756	42,123	253.991
1020	31,020	22,540	235.079	1780	57,473	42,673	254.398
1040	31,688	23,041	235.728	1800	58,191	43,225	254.797
1060	32,357	23,544	236.364	1820	58,910	43,778	255.194
1080	33,029	24,049	236.992	1840	59,629	44,331	255.587
1100	33,702	24,557	237.609	1860	60,351	44,886	255.976
1120	34,377	25,065	238.217	1880	61,072	45,441	256.361
1140	35,054	25,575	238.817	1900	61,794	45,997	256.743
1160	35,733	26,088	239.407	1920	62,516	46,552	257.122
1180	36,406	26,602	239.989	1940	63,238	47,108	257.497
1200	37,095	27,118	240.663	1960	63,961	47,665	257.868
1220	37,780	27,637	241.128	1980	64,684	48,221	258.236
1240	38,466	28,426	241.686	2000	65,408	48,780	258.600
1260	39,154	28,678	242.236	2050	67,224	50,179	259.494
1280	39,844	29,201	242.780	2100	69,044	51,584	260.370
1300	40,534	29,725	243.316	2150	70,864	52,988	261.226
1320	41,226	30,251	243.844	2200	72,688	54,396	262.065
1340	41,919	30,778	244.366	2250	74,516	55,809	262.887
1360	42,613	31,306	244.880	2300	76,345	57,222	263.692
1380	43,309	31,836	245.388	2350	78,178	58,640	264.480
1400	44,007	32,367	245.889	2400	80,015	60,060	265.253
1420	44,707	32,900	246.385	2450	81,852	61,482	266.012
1440	45,408	33,434	246.876	2500	83,692	62,906	266.755
1460	46,110	33,971	247.360	2550	85,537	64,335	267.485
1480	46,813	34,508	247.839	2600	87,383	65,766	268.202
1500	47,517	35,046	248.312	2650	89,230	67,197	268.905
1520	48,222	35,584	248.778	2700	91,077	68,628	269.596
1540	48,928	36,124	249.240	2750	92,930	70,066	270.285
1560	49,635	36,665	249.695	1800	94,784	71,504	270.943
1580	50,344	37,207	250.147	2850	96,639	72,945	271.602
1600	51,053	37,750	250.592	2900	98,495	74,383	272.249
1620	51,763	38,293	251.033	2950	100,352	75,825	272.884
1640	52,472	38,837	251.470	3000	102,210	77,267	273.508
1660	53,184	39,382	251.901	3050	104,073	78,715	274.123
1680	53,895	39,927	252.329	3100	105,939	80,164	274.730
1700	54,609	40,474	252.751	3150	107,802	81,612	275.326
1720	55,323	41,023	253.169	3200	109,667	83,061	275.914
1740	56,039	41,572	253.582	3250	111,534	84,513	276.494

TABLE A-22
Ideal-gas properties of hydrogen, H_2

T K	$\bar{h}$ kJ/kmol	$\bar{u}$ kJ/kmol	$\bar{s}°$ kJ/(kmol · K)	T K	$\bar{h}$ kJ/kmol	$\bar{u}$ kJ/kmol	$\bar{s}°$ kJ/(kmol · K)
0	0	0	0	1440	42,808	30,835	177.410
260	7,370	5,209	126.636	1480	44,091	31,786	178.291
270	7,657	5,412	127.719	1520	45,384	32,746	179.153
280	7,945	5,617	128.765	1560	46,683	33,713	179.995
290	8,233	5,822	129.775	1600	47,990	34,687	180.820
298	8,468	5,989	130.574	1640	49,303	35,668	181.632
300	8,522	6,027	130.754	1680	50,622	36,654	182.428
320	9,100	6,440	132.621	1720	51,947	37,646	183.208
340	9,680	6,853	134.378	1760	53,279	38,645	183.973
360	10,262	7,268	136.039	1800	54,618	39,652	184.724
380	10,843	7,684	137.612	1840	55,962	40,663	185.463
400	11,426	8,100	139.106	1880	57,311	41,680	186.190
420	12,010	8,518	140.529	1920	58,668	42,705	186.904
440	12,594	8,936	141.888	1960	60,031	43,735	187.607
460	13,179	9,355	143.187	2000	61,400	44,771	188.297
480	13,764	9,773	144.432	2050	63,119	46,074	189.148
500	14,350	10,193	145.628	2100	64,847	47,386	189.979
520	14,935	10,611	146.775	2150	66,584	48,708	190.796
560	16,107	11,451	148.945	2200	68,328	50,037	191.598
600	17,280	12,291	150.968	2250	70,080	51,373	192.385
640	18,453	13,133	152.863	2300	71,839	52,716	193.159
680	19,630	13,976	154.645	2350	73,608	54,069	193.921
720	20,807	14,821	156.328	2400	75,383	55,429	194.669
760	21,988	15,669	157.923	2450	77,168	56,798	195.403
800	23,171	16,520	159.440	2500	78,960	58,175	196.125
840	24,359	17,375	160.891	2550	80,755	59,554	196.837
880	25,551	18,235	162.277	2600	82,558	60,941	197.539
920	26,747	19,098	163.607	2650	84,368	62,335	198.229
960	27,948	19,966	164.884	2700	86,186	63,737	198.907
1000	29,154	20,839	166.114	2750	88,008	65,144	199.575
1040	30,364	21,717	167.300	2800	89,838	66,558	200.234
1080	31,580	22,601	168.449	2850	91,671	67,976	200.885
1120	32,802	23,490	169.560	2900	93,512	69,401	201.527
1160	34,028	24,384	170.636	2950	95,358	70,831	202.157
1200	35,262	25,284	171.682	3000	97,211	72,268	202.778
1240	36,502	26,192	172.698	3050	99,065	73,707	203.391
1280	37,749	27,106	173.687	3100	100,926	75,152	203.995
1320	39,002	28,027	174.652	3150	102,793	76,604	204.592
1360	40,263	28,955	175.593	3200	104,667	78,061	205.181
1400	41,530	29,889	176.510	3250	106,545	79,523	205.765

T K	$\bar{h}$ kJ/kmol	$\bar{u}$ kJ/kmol	$\bar{s}°$ kJ/(kmol · K)	T K	$\bar{h}$ kJ/kmol	$\bar{u}$ kJ/kmol	$\bar{s}°$ kJ/(kmol · K)
0	0	0	0	600	20,402	15,413	212.920
220	7,295	5,466	178.576	610	20,765	15,693	213.529
230	7,628	5,715	180.054	620	21,130	15,975	214.122
240	7,961	5,965	181.471	630	21,495	16,257	214.707
250	8,294	6,215	182.831	640	21,862	16,541	215.285
260	8,627	6,466	184.139	650	22,230	16,826	215.856
270	8,961	6,716	185.399	660	22,600	17,112	216.419
280	9,296	6,968	186.616	670	22,970	17,399	216.976
290	9,631	7,219	187.791	680	23,342	17,688	217.527
298	9,904	7,425	188.720	690	23,714	17,978	218.071
300	9,966	7,472	188.928	700	24,088	18,268	218.610
310	10,302	7,725	190.030	710	24,464	18,561	219.142
320	10,639	7,978	191.098	720	24,840	18,854	219.668
330	10,976	8,232	192.136	730	25,218	19,148	220.189
340	11,314	8,487	193.144	740	25,597	19,444	220.707
350	11,652	8,742	194.125	750	25,977	19,741	221.215
360	11,992	8,998	195.081	760	26,358	20,039	221.720
370	12,331	9,255	196.012	770	26,741	20,339	222.221
380	12,672	9,513	196.920	780	27,125	20,639	222.717
390	13,014	9,771	197.807	790	27,510	20,941	223.207
400	13,356	10,030	198.673	800	27,896	21,245	223.693
410	13,699	10,290	199.521	810	28,284	21,549	224.174
420	14,043	10,551	200.350	820	28,672	21,855	224.651
430	14,388	10,813	201.160	830	29,062	22,162	225.123
440	14,734	11,075	201.955	840	29,454	22,470	225.592
450	15,080	11,339	202.734	850	29,846	22,779	226.057
460	15,428	11,603	203.497	860	30,240	23,090	226.517
470	15,777	11,869	204.247	870	30,635	23,402	226.973
480	16,126	12,135	204.982	880	31,032	23,715	227.426
490	16,477	12,403	205.705	890	31,429	24,029	227.875
500	16,828	12,671	206.413	900	31,828	24,345	228.321
510	17,181	12,940	207.112	910	32,228	24,662	228.763
520	17,534	13,211	207.799	920	32,629	24,980	229.202
530	17,889	13,482	208.475	930	33,032	25,300	229.637
540	18,245	13,755	209.139	940	33,436	25,621	230.070
550	18,601	14,028	209.795	950	33,841	25,943	230.499
560	18,959	14,303	210.440	960	34,247	26,265	230.924
570	19,318	14,579	211.075	970	34,653	26,588	231.347
580	19,678	14,856	211.702	980	35,061	26,913	231.767
590	20,039	15,134	212.320	990	35,472	27,240	232.184

TABLE A-23
(Continued)

T K	$\bar{h}$ kJ/kmol	$\bar{u}$ kJ/kmol	$\bar{s}^\circ$ kJ/(kmol · K)	T K	$\bar{h}$ kJ/kmol	$\bar{u}$ kJ/kmol	$\bar{s}^\circ$ kJ/(kmol · K)
1000	35,882	27,568	232.597	1760	70,535	55,902	258.151
1020	36,709	28,228	233.415	1780	71,523	56,723	258.708
1040	37,542	28,895	234.223	1800	72,513	57,547	259.262
1060	38,380	29,567	235.020	1820	73,507	58,375	259.811
1080	39,223	30,243	235.806	1840	74,506	59,207	260.357
1100	40,071	30,925	236.584	1860	75,506	60,042	260.898
1120	40,923	31,611	237.352	1880	76,511	60,880	261.436
1140	41,780	32,301	238.110	1900	77,517	61,720	261.969
1160	42,642	32,997	238.859	1920	78,527	62,564	262.497
1180	43,509	33,698	239.600	1940	79,540	63,411	263.022
1200	44,380	34,403	240.333	1960	80,555	64,259	263.542
1220	45,256	35,112	241.057	1980	81,573	65,111	264.059
1240	46,137	35,827	241.773	2000	82,593	65,965	264.571
1260	47,022	36,546	242.482	2050	85,156	68,111	265.838
1280	47,912	37,270	243.183	2100	87,735	70,275	267.081
1300	48,807	38,000	243.877	2150	90,330	72,454	268.301
1320	49,707	38,732	244.564	2200	92,940	74,649	269.500
1340	50,612	39,470	245.243	2250	95,562	76,855	270.679
1360	51,521	40,213	245.915	2300	98,199	79,076	271.839
1380	52,434	40,960	246.582	2350	100,846	81,308	272.978
1400	53,351	41,711	247.241	2400	103,508	83,553	274.098
1420	54,273	42,466	247.895	2450	106,183	85,811	275.201
1440	55,198	43,226	248.543	2500	108,868	88,082	276.286
1460	56,128	43,989	249.185	2550	111,565	90,364	277.354
1480	57,062	44,756	249.820	2600	114,273	92,656	278.407
1500	57,999	45,528	250.450	2650	116,991	94,958	279.441
1520	58,942	46,304	251.074	2700	119,717	97,269	280.462
1540	59,888	47,084	251.693	2750	122,453	99,588	281.464
1560	60,838	47,868	252.305	2800	125,198	101,917	282.453
1580	61,792	48,655	252.912	2850	127,952	104,256	283.429
1600	62,748	49,445	253.513	2900	130,717	106,605	284.390
1620	63,709	50,240	254.111	2950	133,486	108,959	285.338
1640	64,675	51,039	254.703	3000	136,264	111,321	286.273
1660	65,643	51,841	255.290	3050	139,051	113,692	287.194
1680	66,614	52,646	255.873	3100	141,846	116,072	288.102
1700	67,589	53,455	256.450	3150	144,648	118,458	288.999
1720	68,567	54,267	257.022	3200	147,457	120,851	289.884
1740	69,550	55,083	257.589	3250	150,272	123,250	290.756

T K	$\bar{h}$ kJ/kmol	$\bar{u}$ kJ/kmol	$\bar{s}°$ kJ/(kmol · K)	T K	$\bar{h}$ kJ/kmol	$\bar{u}$ kJ/kmol	$\bar{s}°$ kJ/(kmol · K)
0	0	0	0	2400	50,894	30,940	204.932
298	6,852	4,373	160.944	2450	51,936	31,566	205.362
300	6,892	4,398	161.079	2500	52,979	32,193	205.783
500	11,197	7,040	172.088	2550	54,021	32,820	206.196
1000	21,713	13,398	186.678	2600	55,064	33,447	206.601
1500	32,150	19,679	195.143	2650	56,108	34,075	206.999
1600	34,234	20,931	196.488	2700	57,152	34,703	207.389
1700	36,317	22,183	197.751	2750	58,196	35,332	207.772
1800	38,400	23,434	198.941	2800	59,241	35,961	208.148
1900	40,482	24,685	200.067	2850	60,286	36,590	208.518
2000	42,564	25,935	201.135	2900	61,332	37,220	208.882
2050	43,605	26,560	201.649	2950	62,378	37,851	209.240
2100	44,646	27,186	202.151	3000	63,425	38,482	209.592
2150	45,687	27,811	202.641	3100	65,520	39,746	210.279
2200	46,728	28,436	203.119	3200	67,619	41,013	210.945
2250	47,769	29,062	203.588	3300	69,720	42,283	211.592
2300	48,811	29,688	204.045	3400	71,824	43,556	212.220
2350	49,852	30,314	204.493	3500	73,932	44,832	212.831

T K	$\bar{h}$ kJ/kmol	$\bar{u}$ kJ/kmol	$\bar{s}°$ kJ/(kmol · K)	T K	$\bar{h}$ kJ/kmol	$\bar{u}$ kJ/kmol	$\bar{s}°$ kJ/(kmol · K)
0	0	0	0	2400	77,015	57,061	248.628
298	9,188	6,709	183.594	2450	78,801	58,431	249.364
300	9,244	6,749	183.779	2500	80,592	59,806	250.088
500	15,181	11,024	198.955	2550	82,388	61,186	250.799
1000	30,123	21,809	219.624	2600	84,189	62,572	251.499
1500	46,046	33,575	232.506	2650	85,995	63,962	252.187
1600	49,358	36,055	234.642	2700	87,806	65,358	252.864
1700	52,706	38,571	236.672	2750	89,622	66,757	253.530
1800	56,089	41,123	238.606	2800	91,442	68,162	254.186
1900	59,505	43,708	240.453	2850	93,266	69,570	254.832
2000	62,952	46,323	242.221	2900	95,095	70,983	255.468
2050	64,687	47,642	243.077	2950	96,927	72,400	256.094
2100	66,428	48,968	243.917	3000	98,763	73,820	256.712
2150	68,177	50,301	244.740	3100	102,447	76,673	257.919
2200	69,932	51,641	245.547	3200	106,145	79,539	259.093
2250	71,694	52,987	246.338	3300	109,855	82,418	260.235
2300	73,462	54,339	247.116	3400	113,578	85,309	261.347
2350	75,236	55,697	247.879	3500	117,312	88,212	262.429

TABLE A-26
Enthalpy of formation, Gibbs function of formation, and absolute entropy at 25°C, 1 atm

Substance	Formula	$\bar{h}_f^\circ$ kJ/kmol	$\bar{g}_f^\circ$ kJ/kmol	$\bar{s}^\circ$ kJ/(kmol · K)
Carbon	C(s)	0	0	5.74
Hydrogen	H$_2$(g)	0	0	130.68
Nitrogen	N$_2$(g)	0	0	191.61
Oxygen	O$_2$(g)	0	0	205.04
Carbon monoxide	CO(g)	−110,530	−137,150	197.65
Carbon dioxide	CO$_2$(g)	−393,520	−394,360	213.80
Water vapor	H$_2$O(g)	−241,820	−228,590	188.83
Water	H$_2$O(l)	−285,830	−237,180	69.92
Hydrogen peroxide	H$_2$O$_2$(g)	−136,310	−105,600	232.63
Ammonia	NH$_3$(g)	−46,190	−16,590	192.33
Methane	CH$_4$(g)	−74,850	−50,790	186.16
Acetylene	C$_2$H$_2$(g)	+226,730	+209,170	200.85
Ethylene	C$_2$H$_4$(g)	+52,280	+68,120	219.83
Ethane	C$_2$H$_6$(g)	−84,680	−32,890	229.49
Propylene	C$_3$H$_6$(g)	+20,410	+62,720	266.94
Propane	C$_3$H$_8$(g)	−103,850	−23,490	269.91
n-Butane	C$_4$H$_{10}$(g)	−126,150	−15,710	310.12
n-Octane	C$_8$H$_{18}$(g)	−208,450	+16,530	466.73
n-Octane	C$_8$H$_{18}$(l)	−249,950	+6,610	360.79
n-Dodecane	C$_{12}$H$_{26}$(g)	−291,010	+50,150	622.83
Benzene	C$_6$H$_6$(g)	+82,930	+129,660	269.20
Methyl alcohol	CH$_3$OH(g)	−200,670	−162,000	239.70
Methyl alcohol	CH$_3$OH(l)	−238,660	−166,360	126.80
Ethyl alcohol	C$_2$H$_5$OH(g)	−235,310	−168,570	282.59
Ethyl alcohol	C$_2$H$_5$OH(l)	−277,690	−174,890	160.70
Oxygen	O(g)	+249,190	+231,770	161.06
Hydrogen	H(g)	+218,000	+203,290	114.72
Nitrogen	N(g)	+472,650	+455,510	153.30
Hydroxyl	OH(g)	+39,460	+34,280	183.70

Source: From JANAF, *Thermochemical Tables,* Dow Chemical Co., 1971; *Selected Values of Chemical Thermodynamic Properties,* NBS Technical Note 270-3, 1968; and *API Research Project 44,* Carnegie Press, 1953.

Enthalpy of combustion and enthalpy of vaporization at 25°C, 1 atm
(Water appears as a liquid in the products of combustion)

Substance	Formula	$\Delta \bar{h}_c^\circ = -\textbf{HHV}$ kJ/kmol	$\bar{h}_{fg}$ kJ/kmol
Hydrogen	$H_2(g)$	−285,840	
Carbon	$C(s)$	−393,520	
Carbon monoxide	$CO(g)$	−282,990	
Methane	$CH_4(g)$	−890,360	
Acetylene	$C_2H_2(g)$	−1,299,600	
Ethylene	$C_2H_4(g)$	−1,410,970	
Ethane	$C_2H_6(g)$	−1,559,900	
Propylene	$C_3H_6(g)$	−2,058,500	
Propane	$C_3H_8(g)$	−2,220,000	15,060
n-Butane	$C_4H_{10}(g)$	−2,877,100	21,060
n-Pentane	$C_5H_{12}(g)$	−3,536,100	26,410
n-Hexane	$C_6H_{14}(g)$	−4,194,800	31,530
n-Heptane	$C_7H_{16}(g)$	−4,853,500	36,520
n-Octane	$C_8H_{18}(g)$	−5,512,200	41,460
Benzene	$C_6H_6(g)$	−3,301,500	33,830
Toluene	$C_7H_8(g)$	−3,947,900	39,920
Methyl alcohol	$CH_3OH(g)$	−764,540	37,900
Ethyl alcohol	$C_2H_5OH(g)$	−1,409,300	42,340

Source: Kenneth Wark, *Thermodynamics,* 3d ed., McGraw-Hill, New York, 1977, pp. 834–835, table A-23M.

TABLE A-28

Logarithms to base e of the equilibrium constant K_p

The equilibrium constant K_p for the reaction $v_A + v_B B \rightleftharpoons v_C C + v_D D$ is defined as $K_p \equiv \dfrac{P_C^{v_C} P_D^{v_D}}{P_A^{v_A} P_B^{v_B}}$

Temp. K	$H_2 \rightleftharpoons 2H$	$O_2 \rightleftharpoons 2O$	$N_2 \rightleftharpoons 2N$	$H_2O \rightleftharpoons H_2 + \frac{1}{2}O_2$	$H_2O \rightleftharpoons \frac{1}{2}H_2 + OH$	$CO_2 \rightleftharpoons CO + \frac{1}{2}O_2$	$\frac{1}{2}N_2 + \frac{1}{2}O_2 \rightleftharpoons NO$
298	−164.005	−186.975	−367.480	−92.208	−106.208	−103.762	−35.052
500	−92.827	−105.630	−213.372	−52.691	−60.281	−57.616	−20.295
1000	−39.803	−45.150	−99.127	−23.163	−26.034	−23.529	−9.388
1200	−30.874	−35.005	−80.011	−18.182	−20.283	−17.871	−7.569
1400	−24.463	−27.742	−66.329	−14.609	−16.099	−13.842	−6.270
1600	−19.637	−22.285	−56.055	−11.921	−13.066	−10.830	−5.294
1800	−15.866	−18.030	−48.051	−9.826	−10.657	−8.497	−4.536
2000	−12.840	−14.622	−41.645	−8.145	−8.728	−6.635	−3.931
2200	−10.353	−11.827	−36.391	−6.768	−7.148	−5.120	−3.433
2400	−8.276	−9.497	−32.011	−5.619	−5.832	−3.860	−3.019
2600	−6.517	−7.521	−28.304	−4.648	−4.719	−2.801	−2.671
2800	−5.002	−5.826	−25.117	−3.812	−3.763	−1.894	−2.372
3000	−3.685	−4.357	−22.359	−3.086	−2.937	−1.111	−2.114
3200	−2.534	−3.072	−19.937	−2.451	−2.212	−0.429	−1.888
3400	−1.516	−1.935	−17.800	−1.891	−1.576	0.169	−1.690
3600	−0.609	−0.926	−15.898	−1.392	−1.088	0.701	−1.513
3800	0.202	−0.019	−14.199	−0.945	−0.501	1.176	−1.356
4000	0.934	0.796	−12.660	−0.542	−0.044	1.599	−1.216
4500	2.486	2.513	−9.414	0.312	0.920	2.490	−0.921
5000	3.725	3.895	−6.807	0.996	1.689	3.197	−0.686
5500	4.743	5.023	−4.666	1.560	2.318	3.771	−0.497
6000	5.590	5.963	−2.865	2.032	2.843	4.245	−0.341

Source: Gordon J. Van Wylen and Richard E. Sonntag, *Fundamentals of Classical Thermodynamics,* English/SI Version, 3d ed., Wiley, New York, 1986, p. 723, table A.14. Based on thermodynamic data given in JANAF, *Thermochemical Tables,* Thermal Research Laboratory, The Dow Chemical Company, Midland, Mich.

Constants that appear in the Beattie-Bridgeman and the Benedict-Webb-Rubin equations of state

(a) The Beattie-Bridgeman equation of state is

$$P = \frac{R_u T}{\bar{v}^2}\left(1 - \frac{c}{\bar{v}T^3}\right)(\bar{v} + B) - \frac{A}{\bar{v}^2} \quad \text{where } A = A_0\left(1 - \frac{a}{\bar{v}}\right) \quad \text{and} \quad B = B_0\left(1 - \frac{b}{\bar{v}}\right)$$

When P is in kPa, $\bar{v}$ is in $m^3/kmol$, T is in K, and $R_u = 8.314 \, kPa \cdot m^3/(kmol \cdot K)$, the five constants in the Beattie-Bridgeman equation are as follows:

Gas	A_0	a	B_0	b	c
Air	131.8441	0.019 31	0.046 11	−0.001 101	4.34×10^4
Argon, Ar	130.7802	0.023 28	0.039 31	0.0	5.99×10^4
Carbon dioxide, CO_2	507.2836	0.071 32	0.104 76	0.072 35	6.60×10^5
Helium, He	2.1886	0.059 84	0.014 00	0.0	40
Hydrogen, H_2	20.0117	−0.005 06	0.020 96	−0.043 59	504
Nitrogen, N_2	136.2315	0.026 17	0.050 46	−0.006 91	4.20×10^4
Oxygen, O_2	151.0857	0.025 62	0.046 24	0.004 208	4.80×10^4

Source: Gordon J. Van Wylen and Richard E. Sonntag, *Fundamentals of of Classical Thermodynamics,* English/SI Version, 3d ed., Wiley, New York, 1986, p. 46, table 3.3.

(b) The Benedict-Webb-Rubin equation of state is

$$P = \frac{R_u T}{\bar{v}} + \left(B_0 R_u T - A_0 - \frac{C_0}{T^2}\right)\frac{1}{\bar{v}^2} + \frac{bR_u T - a}{\bar{v}^3} + \frac{a\alpha}{\bar{v}^6} + \frac{c}{\bar{v}^3 T^2}\left(1 + \frac{\gamma}{\bar{v}^2}\right)e^{-\gamma/\bar{v}^2}$$

When P is in kPa, $\bar{v}$ is in $m^3/kmol$, T is in K, and $R_u = 8.314 \, kPa \cdot m^3/(kmol \cdot K)$, the eight constants in the Benedict-Webb-Rubin equation are as follows:

Gas	a	A_0	b	B_0	c	C_0	α	γ
n-Butane, C_4H_{10}	190.68	1021.6	0.039 998	0.124 36	3.205×10^7	1.006×10^8	1.101×10^{-3}	0.0340
Carbon dioxide, CO_2	13.86	277.30	0.007 210	0.049 91	1.511×10^6	1.404×10^7	8.470×10^{-5}	0.00539
Carbon monoxide, CO	3.71	135.87	0.002 632	0.054 54	1.054×10^5	8.673×10^5	1.350×10^{-4}	0.0060
Methane, CH_4	5.00	187.91	0.003 380	0.042 60	2.578×10^5	2.286×10^6	1.244×10^{-4}	0.0060
Nitrogen, N_2	2.54	106.73	0.002 328	0.040 74	7.379×10^4	8.164×10^5	1.272×10^{-4}	0.0053

Source: Kenneth Wark, *Thermodynamics,* 4th ed., McGraw-Hill, New York, 1983, p. 815, table A-21M. Originally published in H. W. Cooper and J. C. Goldfrank, *Hydrocarbon Processing,* vol. 46, p. 141, 1967.

FIGURE A-30a

Nelson–Obert generalized compressibility chart—*low pressures.* (Used with permission of Dr. Edward E. Obert, University of Wisconsin.)

(a) $0 < P_r < 1.0$

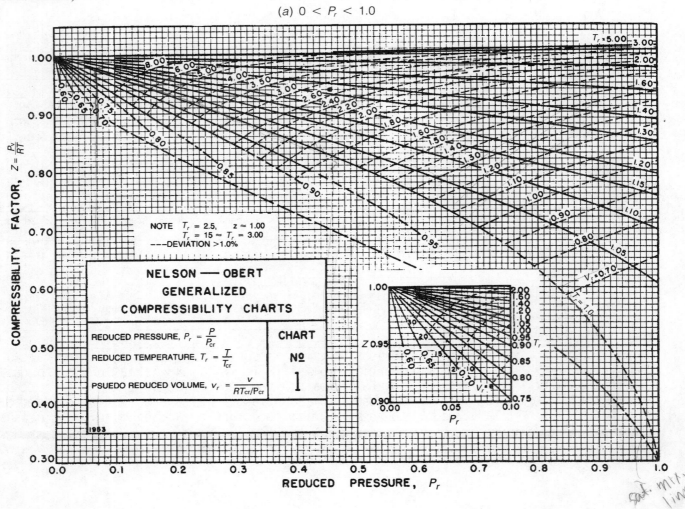

NOTE $T_r = 2.5$, $z \approx 1.00$
$T_r = 15 \approx T_r = 3.00$
----DEVIATION >1.0%

NELSON — OBERT
GENERALIZED
COMPRESSIBILITY CHARTS

REDUCED PRESSURE, $P_r = \dfrac{P}{P_{cr}}$

REDUCED TEMPERATURE, $T_r = \dfrac{T}{T_{cr}}$

PSUEDO REDUCED VOLUME, $v_r = \dfrac{v}{RT_{cr}/P_{cr}}$

CHART
Nº
1

1953

COMPRESSIBILITY FACTOR, $Z = \dfrac{Pv}{RT}$

REDUCED PRESSURE, P_r

916

Nelson–Obert generalized compressibility chart—*intermediate pressures.* (Used with permission of Dr. Edward E. Obert, University of Wisconsin.)

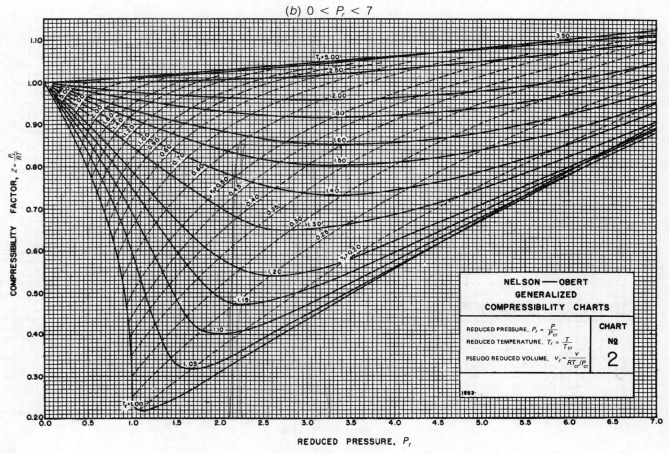

(b) $0 < P_r < 7$

REDUCED PRESSURE, P_r

Nelson–Obert generalized compressibility chart—*high pressures.* (Used with permission of Dr. Edward E. Obert, University of Wisconsin.)

(*c*) 0 < P_r < 40

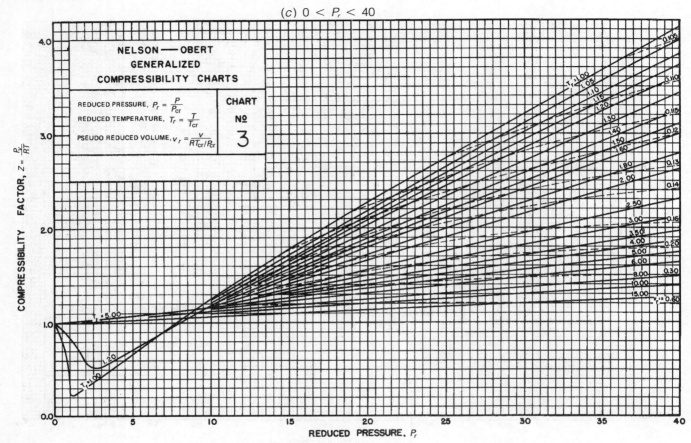

Generalized enthalpy departure chart. (*Source:* John R. Howell and Richard O. Buckius. *Fundamentals of Engineering Thermodynamics,* SI Version, McGraw-Hill, New York, 1987, p. 558, fig. C.2, and p. 561, fig. C.5.)

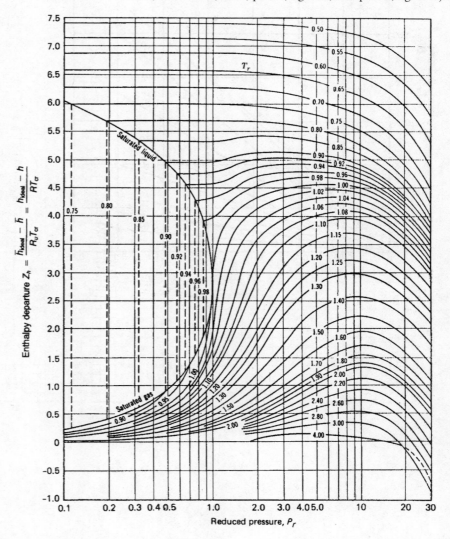

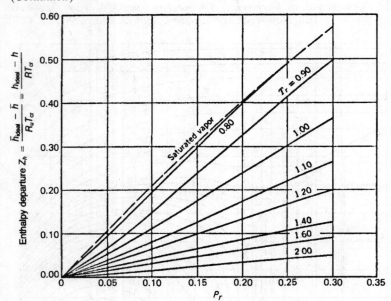

Generalized entropy departure chart. (*Source:* John R. Howell and Richard O. Buckius, *Fundamentals of Engineering Thermodynamics,* SI Version, McGraw-Hill, New York, 1987, p. 559, fig. C.3, and p. 561, fig. C.5)

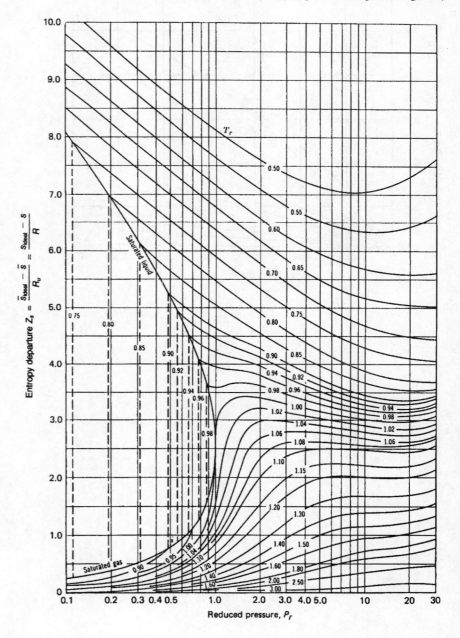

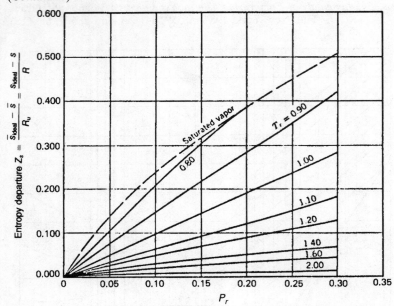

Psychrometric chart at 1-atm total pressure. (Reprinted by permission of the American Society of Heating, Refrigerating and Air-Conditioning Engineers, Inc., Atlanta.)

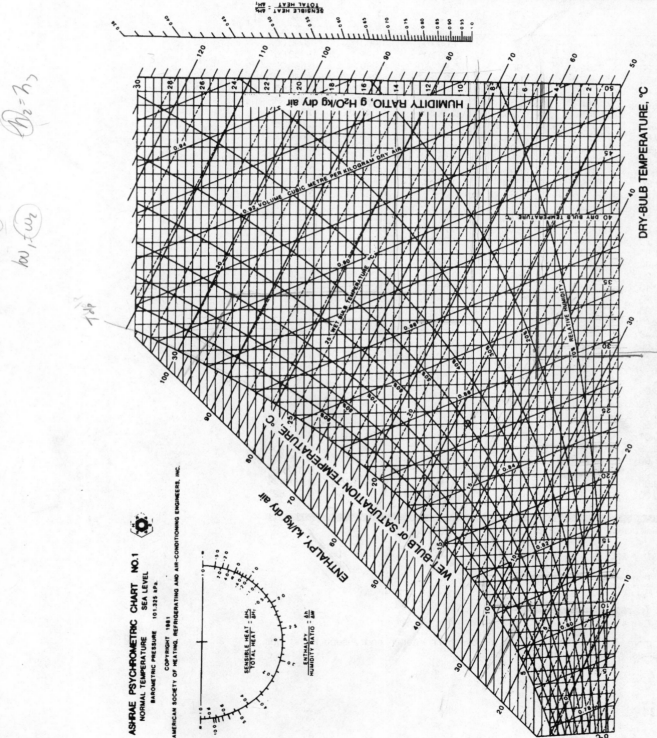

TABLE A-34

One-dimensional isentropic compressible-flow functions for an ideal gas with constant specific heats and molecular weight and $k = 1.4$†

M	M^*	$\dfrac{A}{A^*}$	$\dfrac{P}{P_0}$	$\dfrac{\rho}{\rho_0}$	$\dfrac{T}{T_0}$
0	0	∞	1.000 00	1.000 00	1.000 00
0.10	0.109 43	5.8218	0.993 03	0.995 02	0.998 00
0.20	0.218 22	2.9635	0.972 50	0.980 27	0.992 06
0.30	0.325 72	2.0351	0.939 47	0.956 38	0.982 32
0.40	0.431 33	1.5901	0.895 62	0.924 28	0.968 99
0.50	0.534 52	1.3398	0.843 02	0.885 17	0.952 38
0.60	0.634 80	1.1882	0.784 00	0.840 45	0.932 84
0.70	0.731 79	1.094 37	0.720 92	0.791 58	0.910 75
0.80	0.825 14	1.038 23	0.656 02	0.740 00	0.886 52
0.90	0.914 60	1.008 86	0.591 26	0.687 04	0.860 58
1.00	1.000 00	1.000 00	0.528 28	0.633 94	0.833 33
1.10	1.081 24	1.007 93	0.468 35	0.581 69	0.805 15
1.20	1.1583	1.030 44	0.412 38	0.531 14	0.776 40
1.30	1.2311	1.066 31	0.360 92	0.482 91	0.747 38
1.40	1.2999	1.1149	0.314 24	0.437 42	0.718 39
1.50	1.3646	1.1762	0.272 40	0.394 98	0.689 65
1.60	1.4254	1.2502	0.235 27	0.355 73	0.661 38
1.70	1.4825	1.3376	0.202 59	0.319 69	0.633 72
1.80	1.5360	1.4390	0.174 04	0.286 82	0.606 80
1.90	1.5861	1.5552	0.149 24	0.256 99	0.580 72
2.00	1.6330	1.6875	0.127 80	0.230 05	0.555 56
2.10	1.6769	1.8369	0.109 35	0.205 80	0.531 35
2.20	1.7179	2.0050	0.093 52	0.184 05	0.508 13
2.30	1.7563	2.1931	0.079 97	0.164 58	0.485 91
2.40	1.7922	2.4031	0.068 40	0.147 20	0.464 68
2.50	1.8258	2.6367	0.058 53	0.131 69	0.444 44
2.60	1.8572	2.8960	0.050 12	0.117 87	0.425 17
2.70	1.8865	3.1830	0.042 95	0.105 57	0.406 84
2.80	1.9140	3.5001	0.036 85	0.094 62	0.389 41
2.90	1.9398	3.8498	0.031 65	0.084 89	0.372 86
3.00	1.9640	4.2346	0.027 22	0.076 23	0.357 14
3.50	2.0642	6.7896	0.013 11	0.045 23	0.289 86
4.00	2.1381	10.719	0.006 58	0.027 66	0.238 10
4.50	2.1936	16.562	0.003 46	0.017 45	0.198 02
5.00	2.2361	25.000	$189(10)^{-5}$	0.011 34	0.166 67
6.00	2.2953	53.180	$633(10)^{-6}$	0.005 19	0.121 95
7.00	2.3333	104.143	$242(10)^{-6}$	0.002 61	0.092 59
8.00	2.3591	190.109	$102(10)^{-6}$	0.001 41	0.072 46
9.00	2.3772	327.189	$474(10)^{-7}$	0.000 815	0.058 14
10.00	2.3904	535.938	$236(10)^{-7}$	0.000 495	0.047 62
∞	2.4495	∞	0	0	0

†Calculated from the following relations for $k = 1.4$:

$$M^* = M\sqrt{\frac{k+1}{2+(k-1)M^2}} \qquad \frac{A}{A^*} = \frac{1}{M}\left[\left(\frac{2}{k+1}\right)\left(1+\frac{k-1}{2}M^2\right)\right]^{(k+1)/[2(k-1)]}$$

$$\frac{P}{P_0} = \left(1+\frac{k-1}{2}M^2\right)^{-k/(k-1)} \qquad \frac{\rho}{\rho_0} = \left(1+\frac{k-1}{2}M^2\right)^{-1/(k-1)}$$

$$\frac{T}{T_0} = \left(1+\frac{k-1}{2}M^2\right)^{-1}$$

A more extensive table is given in Joseph H. Keenan and Joseph Kaye, *Gas Tables*, Wiley, New York, 1948, table 30.

One-dimensional normal-shock functions for an ideal gas with constant specific heats and molecular weight and $k = 1.4$†

M_x	M_y	$\dfrac{P_y}{P_x}$	$\dfrac{\rho_y}{\rho_x}$	$\dfrac{T_y}{T_x}$	$\dfrac{P_{0y}}{P_{0x}}$	$\dfrac{P_{0y}}{P_x}$
1.00	1.000 00	1.0000	1.0000	1.0000	1.000 00	1.8929
1.10	0.911 77	1.2450	1.1691	1.0649	0.998 92	2.1328
1.20	0.842 17	1.5133	1.3416	1.1280	0.992 80	2.4075
1.30	0.785 96	1.8050	1.5157	1.1909	0.979 35	2.7135
1.40	0.739 71	2.1200	1.6896	1.2547	0.958 19	3.0493
1.50	0.701 09	2.4583	1.8621	1.3202	0.929 78	3.4133
1.60	0.668 44	2.8201	2.0317	1.3880	0.895 20	3.8049
1.70	0.640 55	3.2050	2.1977	1.4583	0.855 73	4.2238
1.80	0.616 50	3.6133	2.3592	1.5316	0.812 68	4.6695
1.90	0.595 62	4.0450	2.5157	1.6079	0.767 35	5.1417
2.00	0.577 35	4.5000	2.6666	1.6875	0.720 88	5.6405
2.10	0.561 28	4.9784	2.8119	1.7704	0.674 22	6.1655
2.20	0.547 06	5.4800	2.9512	1.8569	0.628 12	6.7163
2.30	0.534 41	6.0050	3.0846	1.9468	0.583 31	7.2937
2.40	0.523 12	6.5533	3.2119	2.0403	0.540 15	7.8969
2.50	0.512 99	7.1250	3.3333	2.1375	0.499 02	8.5262
2.60	0.503 87	7.7200	3.4489	2.2383	0.460 12	9.1813
2.70	0.495 63	8.3383	3.5590	2.3429	0.423 59	9.8625
2.80	0.488 17	8.9800	3.6635	2.4512	0.389 46	10.569
2.90	0.481 38	9.6450	3.7629	2.5632	0.357 73	11.302
3.00	0.475 19	10.333	3.8571	2.6790	0.328 34	12.061
4.00	0.434 96	18.500	4.5714	4.0469	0.138 76	21.068
5.00	0.415 23	29.000	5.0000	5.8000	0.061 72	32.654
10.00	0.387 57	116.50	5.7143	20.388	0.003 04	129.217
∞	0.377 96	∞	6.000	∞	0	∞

†Calculated from the following relations for $k = 1.4$:

$$M_y = \sqrt{\frac{M_x^2 + 2/(k-1)}{2M_x^2 k/(k-1) - 1}} \qquad \frac{T_y}{T_x} = \frac{1 + M_x^2(k-1)/2}{1 + M_y^2(k-1)/2}$$

$$\frac{P_y}{P_x} = \frac{1 + kM_x^2}{1 + kM_y^2} \qquad \frac{P_{0y}}{P_{0x}} = \frac{M_x}{M_y}\left[\frac{1 + M_y^2(k-1)/2}{1 + M_x^2(k-1)/2}\right]^{(k+1)/[2(k-1)]}$$

$$\frac{\rho_y}{\rho_x} = \frac{P_y/P_x}{T_y/T_x} \qquad \frac{P_{0y}}{P_x} = \frac{(1 + kM_x^2)[1 + M_y^2(k-1)/2]^{k/(k-1)}}{1 + kM_y^2}$$

A more extensive table is given in Joseph H. Keenan and Joseph Kaye, *Gas Tables*, Wiley, New York, 1948, table 30.

Property Tables, Figures, and Charts (English Units)

APPENDIX

2

TABLE A-1E
Molar mass, gas constant, and critical-point properties

Substance	Formula	Molar mass lbm/lbmol	Gas constant, R Btu/(lbm · R)*	psia · ft³/(lbm · R)*	Temp. R	Pressure psia	Volume ft³/lbmol
Ammonia	NH_3	17.03	0.1166	0.6301	729.8	1636	1.16
Argon	Ar	39.948	0.049 71	0.2686	272	705	1.20
Bromine	Br_2	159.808	0.012 43	0.067 14	1052	1500	2.17
Carbon dioxide	CO_2	44.01	0.045 13	0.2438	547.5	1071	1.51
Carbon monoxide	CO	28.011	0.070 90	0.3831	240	507	1.49
Chlorine	Cl_2	70.906	0.028 01	0.1517	751	1120	1.99
Deuterium (normal)	D_2	4.00	0.4965	2.6825	69.1	241	—
Helium	He	4.003	0.4961	2.6805	9.5	33.2	0.926
Hydrogen (normal)	H_2	2.016	0.9851	5.3224	59.9	188.1	1.04
Krypton	Kr	83.80	0.023 70	0.1280	376.9	798	1.48
Neon	Ne	20.183	0.098 40	0.5316	80.1	395	0.668
Nitrogen	N_2	28.013	0.070 90	0.3830	227.1	492	1.44
Nitrous oxide	N_2O	44.013	0.045 12	0.2438	557.4	1054	1.54
Oxygen	O_2	31.999	0.062 06	0.3353	278.6	736	1.25
Sulfur dioxide	SO_2	64.063	0.031 00	1.1675	775.2	1143	1.95
Water	H_2O	18.015	0.1102	0.5956	1165.3	3204	0.90
Xenon	Xe	131.30	0.015 13	0.081 72	521.55	852	1.90
Benzene	C_6H_6	78.115	0.025 42	0.1374	1012	714	4.17
n-Butane	C_4H_{10}	58.124	0.034 17	0.1846	765.2	551	4.08
Carbon tetrachloride	CCl_4	153.82	0.012 91	0.069 76	1001.5	661	4.42
Chloroform	$CHCl_3$	119.38	0.016 64	0.089 88	965.8	794	3.85
Dichlorodifluoromethane	CCl_2F_2 (R-12)	120.91	0.016 43	0.088 74	692.4	582	3.49
Dichlorofluoromethane	$CHCl_2F$	102.92	0.019 30	0.1043	813.0	749	3.16
Ethane	C_2H_6	30.020	0.066 16	0.3574	549.8	708	2.37
Ethyl alcohol	C_2H_5OH	46.07	0.043 11	0.2329	929.0	926	2.68
Ethylene	C_2H_4	28.054	0.070 79	0.3825	508.3	742	1.99
n-Hexane	C_6H_{14}	86.178	0.023 05	0.1245	914.2	439	5.89
Methane	CH_4	16.043	0.1238	0.6688	343.9	673	1.59
Methyl alcohol	CH_3OH	32.042	0.061 98	0.3349	923.7	1154	1.89
Methyl chloride	CH_3Cl	50.488	0.039 34	0.2125	749.3	968	2.29
Propane	C_2H_8	44.097	0.045 04	0.2433	665.9	617	3.20
Propene	C_3H_6	42.081	0.047 19	0.2550	656.9	670	2.90
Propyne	C_3H_4	40.065	0.049 57	0.2678	722	776	—
Trichlorofluoromethane	CCl_3F	137.37	0.014 46	0.078 11	848.1	635	3.97
Air	—	28.97	0.068 55	0.3704	—	—	—

*Calculated from $R = R_u/M$, where $R_u = 1.986$ Btu/(lbmol · R) = 10.73 psia · ft³/(lbmol · R) and M is the molar mass.

Source: Gordon J. Van Wylen and Richard E. Sonntag, *Fundamentals of Classical Thermodynamics*, English/SI Version, 3d ed., Wiley, New York, 1986, p. 684, table A.6E. Originally published in K. A. Kobe and R. E. Lynn, Jr., *Chemical Review,* vol. 52, pp. 117–236, 1953.

Gas	Formula	Gas constant R Btu/(lbm · R)	C_{p0} Btu/(lbm · R)	C_{v0} Btu/(lbm · R)	k
Air	—	0.068 55	0.240	0.171	1.400
Argon	Ar	0.049 71	0.1253	0.0756	1.667
Butane	C_4H_{10}	0.034 24	0.415	0.381	1.09
Carbon dioxide	CO_2	0.045 13	0.203	0.158	1.285
Carbon monoxide	CO	0.070 90	0.249	0.178	1.399
Ethane	C_2H_6	0.066 16	0.427	0.361	1.183
Ethylene	C_2H_4	0.070 79	0.411	0.340	1.208
Helium	He	0.4961	1.25	0.753	1.667
Hydrogen	H_2	0.9851	3.43	2.44	1.404
Methane	CH_4	0.1238	0.532	0.403	1.32
Neon	Ne	0.098 40	0.246	0.1477	1.667
Nitrogen	N_2	0.070 90	0.248	0.177	1.400
Octane	C_8H_{18}	0.017 42	0.409	0.392	1.044
Oxygen	O_2	0.062 06	0.219	0.157	1.395
Propane	C_3H_8	0.045 04	0.407	0.362	1.124
Steam	H_2O	0.1102	0.445	0.335	1.329

Source: Gordon J. Van Wylen and Richard E. Sonntag, *Fundamentals of Classical Thermodynamics,* English/SI Version, 3d ed., Wiley, New York, 1986, p. 687, table A-8E.

TABLE A-2E
(*Continued*)

(*b*) At various temperatures

Temp. °F	C_{p0} Btu/(lbm · R)	C_{v0} Btu/(lbm · R)	k	C_{p0} Btu/(lbm · R)	C_{v0} Btu/(lbm · R)	k	C_{p0} Btu/(lbm · R)	C_{v0} Btu/(lbm · R)	k
		Air			Carbon dioxide, CO_2			Carbon monoxide, CO	
40	0.240	0.171	1.401	0.195	0.150	1.300	0.248	0.177	1.400
100	0.240	0.172	1.400	0.205	0.160	1.283	0.249	0.178	1.399
200	0.241	0.173	1.397	0.217	0.172	1.262	0.249	0.179	1.397
300	0.243	0.174	1.394	0.229	0.184	1.246	0.251	0.180	1.394
400	0.245	0.176	1.389	0.239	0.193	1.233	0.253	0.182	1.389
500	0.248	0.179	1.383	0.247	0.202	1.223	0.256	0.185	1.384
600	0.250	0.182	1.377	0.255	0.210	1.215	0.259	0.188	1.377
700	0.254	0.185	1.371	0.262	0.217	1.208	0.262	0.191	1.371
800	0.257	0.188	1.365	0.269	0.224	1.202	0.266	0.195	1.364
900	0.259	0.191	1.358	0.275	0.230	1.197	0.269	0.198	1.357
1000	0.263	0.195	1.353	0.280	0.235	1.192	0.273	0.202	1.351
1500	0.276	0.208	1.330	0.298	0.253	1.178	0.287	0.216	1.328
2000	0.286	0.217	1.312	0.312	0.267	1.169	0.297	0.226	1.314
		Hydrogen, H_2			Nitrogen, N_2			Oxygen, O_2	
40	3.397	2.412	1.409	0.248	0.177	1.400	0.219	0.156	1.397
100	3.426	2.441	1.404	0.248	0.178	1.399	0.220	0.158	1.394
200	3.451	2.466	1.399	0.249	0.178	1.398	0.223	0.161	1.387
300	3.461	2.476	1.398	0.250	0.179	1.396	0.226	0.164	1.378
400	3.466	2.480	1.397	0.251	0.180	1.393	0.230	0.168	1.368
500	3.469	2.484	1.397	0.254	0.183	1.388	0.235	0.173	1.360
600	3.473	2.488	1.396	0.256	0.185	1.383	0.239	0.177	1.352
700	3.477	2.492	1.395	0.260	0.189	1.377	0.242	0.181	1.344
800	3.494	2.509	1.393	0.262	0.191	1.371	0.246	0.184	1.337
900	3.502	2.519	1.392	0.265	0.194	1.364	0.249	0.187	1.331
1000	3.513	2.528	1.390	0.269	0.198	1.359	0.252	0.190	1.326
1500	3.618	2.633	1.374	0.283	0.212	1.334	0.263	0.201	1.309
2000	3.758	2.773	1.355	0.293	0.222	1.319	0.270	0.208	1.298

Source: Kenneth Wark, *Thermodynamics*, 4th ed., McGraw-Hill, New York, 1983, p. 830, table A-4. Originally published in *Tables of Properties of Gases*, NBS *Circular*, 564, 1955.

(*c*) As a function of temperature

$$\bar{C}_{p0} = a + bT + cT^2 + dT^3$$
[T in R, $\bar{C}_{p0}$ in Btu/(lbmol · R)]

Substance	Formula	a	b	c	d	Temperature range R	% error Max.	% error Avg.
Nitrogen	N_2	6.903	$-0.020\,85 \times 10^{-2}$	$0.059\,57 \times 10^{-5}$	-0.1176×10^{-9}	491–3240	0.59	0.34
Oxygen	O_2	6.085	0.2017×10^{-2}	$-0.052\,75 \times 10^{-5}$	$0.053\,72 \times 10^{-9}$	491–3240	1.19	0.28
Air	—	6.713	$0.026\,09 \times 10^{-2}$	$0.035\,40 \times 10^{-5}$	$-0.080\,52 \times 10^{-9}$	491–3240	0.72	0.33
Hydrogen	H_2	6.952	$-0.025\,42 \times 10^{-2}$	$0.029\,52 \times 10^{-5}$	$-0.035\,65 \times 10^{-9}$	491–3240	1.02	0.26
Carbon monoxide	CO	6.726	$0.022\,22 \times 10^{-2}$	$0.039\,60 \times 10^{-5}$	$-0.091\,00 \times 10^{-9}$	491–3240	0.89	0.37
Carbon dioxide	CO_2	5.316	$0.793\,61 \times 10^{-2}$	-0.2581×10^{-5}	0.3059×10^{-9}	491–3240	0.67	0.22
Water vapor	H_2O	7.700	$0.025\,52 \times 10^{-2}$	$0.077\,81 \times 10^{-5}$	-0.1472×10^{-9}	491–3240	0.53	0.24
Nitric oxide	NO	7.008	$-0.012\,47 \times 10^{-2}$	$0.071\,85 \times 10^{-5}$	-0.1715×10^{-9}	491–2700	0.97	0.36
Nitrous oxide	N_2O	5.758	0.7780×10^{-2}	-0.2596×10^{-5}	0.4331×10^{-9}	491–2700	0.59	0.26
Nitrogen dioxide	NO_2	5.48	0.7583×10^{-2}	-0.260×10^{-5}	0.322×10^{-9}	491–2700	0.46	0.18
Ammonia	NH_3	6.5846	$0.340\,28 \times 10^{-2}$	$0.073\,034 \times 10^{-5}$	$-0.274\,02 \times 10^{-9}$	491–2700	0.91	0.36
Sulfur	S_2	6.499	0.2943×10^{-2}	-0.1200×10^{-5}	0.1632×10^{-9}	491–3240	0.99	0.38
Sulfur dioxide	SO_2	6.157	0.7689×10^{-2}	-0.2810×10^{-5}	0.3527×10^{-9}	491–3240	0.45	0.24
Sulfur trioxide	SO_3	3.918	1.935×10^{-2}	-0.8256×10^{-5}	1.328×10^{-9}	491–2340	0.29	0.13
Acetylene	C_2H_2	5.21	1.2227×10^{-2}	-0.4812×10^{-5}	0.7457×10^{-9}	491–2700	1.46	0.59
Benzene	C_6H_6	−8.650	6.4322×10^{-2}	-2.327×10^{-5}	3.179×10^{-9}	491–2700	0.34	0.20
Methanol	CH_4O	4.55	1.214×10^{-2}	-0.0898×10^{-5}	-0.329×10^{-9}	491–1800	0.18	0.08
Ethanol	C_2H_6O	4.75	2.781×10^{-2}	-0.7651×10^{-5}	0.821×10^{-9}	491–2700	0.40	0.22
Hydrogen chloride	HCl	7.244	-0.1011×10^{-2}	$0.097\,83 \times 10^{-5}$	-0.1776×10^{-9}	491–2740	0.22	0.08
Methane	CH_4	4.750	0.6666×10^{-2}	$0.093\,52 \times 10^{-5}$	-0.4510×10^{-9}	491–2740	1.33	0.57
Ethane	C_2H_6	1.648	2.291×10^{-2}	-0.4722×10^{-5}	0.2984×10^{-9}	491–2740	0.83	0.28
Propane	C_3H_8	−0.966	4.044×10^{-2}	-1.159×10^{-5}	1.300×10^{-9}	491–2740	0.40	0.12
n-Butane	C_4H_{10}	0.945	4.929×10^{-2}	-1.352×10^{-5}	1.433×10^{-9}	491–2740	0.54	0.24
i-Butane	C_4H_{10}	−1.890	5.520×10^{-2}	-1.696×10^{-5}	2.044×10^{-9}	491–2740	0.25	0.13
n-Pentane	C_5H_{12}	1.618	6.028×10^{-2}	-1.656×10^{-5}	1.732×10^{-9}	491–2740	0.56	0.21
n-Hexane	C_6H_{14}	1.657	7.328×10^{-2}	-2.112×10^{-5}	2.363×10^{-9}	491–2740	0.72	0.20
Ethylene	C_2H_4	0.944	2.075×10^{-2}	-0.6151×10^{-5}	0.7326×10^{-9}	491–2740	0.54	0.13
Propylene	C_3H_6	0.753	3.162×10^{-2}	-0.8981×10^{-5}	1.008×10^{-9}	491–2740	0.73	0.17

Source: B. G. Kyle, *Chemical and Process Thermodynamics*, Prentice-Hall, Englewood Cliffs, N.J., 1984. Used with permission.

TABLE A-3E
Specific heats and densities of common solids and liquids

(*a*) At 80°F

Solid	C_p Btu/(lbm · R)	ρ lbm/ft³	Liquid	C_p Btu/(lbm · R)	ρ lbm/ft³
Aluminum	0.215	170	Ammonia	1.146	38
Copper	0.092	555	Ethanol	0.587	49
Granite	0.243	170	Refrigerant-12	0.233	82
Graphite	0.170	155	Mercury	0.033	847
Iron	0.107	490	Methanol	0.609	49
Lead	0.030	705	Oil (light)	0.430	57
Rubber (soft)	0.439	70	Water	1.000	62
Silver	0.056	655			
Tin	0.052	360			
Wood (most)	0.420	22–45			

Source: Gordon J. Van Wylen and Richard E. Sonntag, *Fundamentals of Classical Thermodynamics,* English/SI Version, 3d ed., Wiley, New York, 1986, p. 686, table A-7E.

(*b*) At various temperatures

Solids

Substance	Temp. °F	C_p Btu/(lbm · R)	Substance	Temp. °F	C_p Btu/(lbm · R)
Ice	−100	0.375	Lead	−455	0.0008
	−50	0.424		−435	0.0073
	0	0.471		−150	0.0283
	20	0.491		32	0.0297
	32	0.502		210	0.0320
Aluminum	−150	0.167		570	0.0356
	−100	0.192	Copper	−240	0.0674
	32	0.212		−150	0.0784
	100	0.218		−60	0.0862
	200	0.224		0	0.0893
	300	0.229		100	0.0925
	400	0.235		200	0.0938
	500	0.240		390	0.0963
Iron	68	0.107	Silver	68	0.0558

Liquids

Substance	State	C_p Btu/(lbm · R)	Substance	State	C_p Btu/(lbm · R)
Water	1 atm, 32°F	1.007	Glycerin	1 atm, 50°F	0.554
	1 atm, 77°F	0.998		1 atm, 120°F	0.617
	1 atm, 212°F	1.007	Bismuth	1 atm, 800°F	0.0345
Ammonia	Sat., 0°F	1.08		1 atm, 1400°F	0.0393
	Sat., 120°F	1.22	Mercury	1 atm, 50°F	0.033
Refrigerant-12	Sat., −40°F	0.211		1 atm, 600°F	0.032
	Sat., 0°F	0.217	Sodium	1 atm, 200°F	0.33
	Sat., 120°F	0.244		1 atm, 1000°F	0.30
Benzene	1 atm, 60°F	0.43	Propane	1 atm, 32°F	0.576
	1 atm, 150°F	0.46			

Source: Kenneth Wark, *Thermodynamics,* 4th ed., McGraw-Hill, New York, 1983, p. 862, table A-19.

Temp. °F T	Sat. press. psia P_{sat}	Specific volume ft³/lbm Sat. liquid v_f	Sat. vapor v_g	Internal energy Btu/lbm Sat. liquid u_f	Evap. u_{fg}	Sat. vapor u_g	Enthalpy Btu/lbm Sat. liquid h_f	Evap. h_{fg}	Sat. vapor h_g	Entropy Btu/(lbm·R) Sat. liquid s_f	Evap. s_{fg}	Sat. vapor s_g
32.018	0.088 66	0.016 022	3302	0.00	1021.2	1021.2	0.01	1075.4	1075.4	0.000 00	2.1869	2.1869
35	0.099 92	0.016 021	2948	2.99	1019.2	1022.2	3.00	1073.7	1076.7	0.006 07	2.1704	2.1764
40	0.121 66	0.016 020	2445	8.02	1015.8	1023.9	8.02	1070.9	1078.9	0.016 17	2.1430	2.1592
45	0.147 48	0.016 021	2037	13.04	1012.5	1025.5	13.04	1068.1	1081.1	0.026 18	2.1162	2.1423
50	0.178 03	0.016 024	1704.2	18.06	1009.1	1027.2	18.06	1065.2	1083.3	0.036 07	2.0899	2.1259
60	0.2563	0.016 035	1206.9	28.08	1002.4	1030.4	28.08	1059.6	1087.7	0.055 55	2.0388	2.0943
70	0.3632	0.016 051	867.7	38.09	995.6	1033.7	38.09	1054.0	1092.0	0.074 63	1.9896	2.0642
80	0.5073	0.016 073	632.8	48.08	988.9	1037.0	48.09	1048.3	1096.4	0.093 32	1.9423	2.0356
90	0.6988	0.016 099	467.7	58.07	982.2	1040.2	58.07	1042.7	1100.7	0.111 65	1.8966	2.0083
100	0.9503	0.016 130	350.0	68.04	975.4	1043.5	68.05	1037.0	1105.0	0.129 63	1.8526	1.9822
110	1.2763	0.016 166	265.1	78.02	968.7	1046.7	78.02	1031.3	1109.3	0.147 30	1.8101	1.9574
120	1.6945	0.016 205	203.0	87.99	961.9	1049.9	88.00	1025.5	1113.5	0.164 65	1.7690	1.9336
130	2.225	0.016 247	157.17	97.97	955.1	1053.0	97.98	1019.8	1117.8	0.181 72	1.7292	1.9109
140	2.892	0.016 293	122.88	107.95	948.2	1056.2	107.96	1014.0	1121.9	0.198 51	1.6907	1.8892
150	3.722	0.016 343	96.99	117.95	941.3	1059.3	117.96	1008.1	1126.1	0.215 03	1.6533	1.8684
160	4.745	0.016 395	77.23	127.94	934.4	1062.3	127.96	1002.2	1130.1	0.231 30	1.6171	1.8484
170	5.996	0.016 450	62.02	137.95	927.4	1065.4	137.97	996.2	1134.2	0.247 32	1.5819	1.8293
180	7.515	0.016 509	50.20	147.97	920.4	1068.3	147.99	990.2	1138.2	0.263 11	1.5478	1.8109
190	9.343	0.016 570	40.95	158.00	913.3	1071.3	158.03	984.1	1142.1	0.278 66	1.5146	1.7932
200	11.529	0.016 634	33.63	168.04	906.2	1074.2	168.07	977.9	1145.9	0.294 00	1.4822	1.7762
210	14.125	0.016 702	27.82	178.10	898.9	1077.0	178.14	971.6	1149.7	0.309 13	1.4508	1.7599
212	14.698	0.016 716	26.80	180.11	897.5	1077.6	180.16	970.3	1150.5	0.312 13	1.4446	1.7567
220	17.188	0.016 772	23.15	188.17	891.7	1079.8	188.22	965.3	1153.5	0.324 06	1.4201	1.7441
230	20.78	0.016 845	19.386	198.26	884.3	1082.6	198.32	958.8	1157.1	0.338 80	1.3901	1.7289
240	24.97	0.016 922	16.327	208.36	876.9	1085.3	208.44	952.3	1160.7	0.353 35	1.3609	1.7143
250	29.82	0.017 001	13.826	218.49	869.4	1087.9	218.59	945.6	1164.2	0.367 72	1.3324	1.7001
260	35.42	0.017 084	11.768	228.64	861.8	1090.5	228.76	938.8	1167.6	0.381 93	1.3044	1.6864
270	41.85	0.017 170	10.066	238.82	854.1	1093.0	238.95	932.0	1170.9	0.395 97	1.2771	1.6731
280	49.18	0.017 259	8.650	249.02	846.3	1095.4	249.18	924.9	1174.1	0.409 86	1.2504	1.6602
290	57.53	0.017 352	7.467	259.25	838.5	1097.7	259.44	917.8	1177.2	0.423 60	1.2241	1.6477
300	66.98	0.017 448	6.472	269.52	830.5	1100.0	269.73	910.4	1180.2	0.437 20	1.1984	1.6356
310	77.64	0.017 548	5.632	279.81	822.3	1102.1	280.06	903.0	1183.0	0.450 67	1.1731	1.6238
320	89.60	0.017 652	4.919	290.14	814.1	1104.2	290.43	895.3	1185.8	0.464 00	1.1483	1.6123
330	103.00	0.017 760	4.312	300.51	805.7	1106.2	300.43	887.5	1188.4	0.477 22	1.1238	1.6010
340	117.93	0.017 872	3.792	310.91	797.1	1108.0	311.30	879.5	1190.8	0.490 31	1.0997	1.5901
350	134.53	0.017 988	3.346	321.35	788.4	1109.8	321.80	871.3	1193.1	0.503 29	1.0760	1.5793
360	152.92	0.018 108	2.961	331.84	779.6	1111.4	332.35	862.9	1195.2	0.516 17	1.0526	1.5688
370	173.23	0.018 233	2.628	342.37	770.6	1112.9	342.96	854.2	1197.2	0.528 94	1.0295	1.5585
380	195.60	0.018 363	2.339	352.95	761.4	1114.3	353.62	845.4	1199.0	0.541 63	1.0067	1.5483
390	220.2	0.018 498	2.087	363.58	752.0	1115.6	364.34	836.2	1200.6	0.554 22	0.9841	1.5383
400	247.1	0.018 638	1.8661	374.27	742.4	1116.6	375.12	826.8	1202.0	0.566 72	0.9617	1.5284
410	276.5	0.018 784	1.6726	385.01	732.6	1117.6	385.97	817.2	1203.1	0.579 16	0.9395	1.5187
420	308.5	0.018 936	1.5024	395.81	722.5	1118.3	396.89	807.2	1204.1	0.591 52	0.9175	1.5091
430	343.3	0.019 094	1.3521	406.68	712.2	1118.9	407.89	796.9	1204.8	0.603 81	0.8957	1.4995
440	381.2	0.019 260	1.2192	417.62	701.7	1119.3	418.98	786.3	1205.3	0.616 05	0.8740	1.4900
450	422.1	0.019 433	1.1011	428.6	690.9	1119.5	430.2	775.4	1205.6	0.6282	0.8523	1.4806
460	466.3	0.019 614	0.9961	439.7	679.8	1119.6	441.4	764.1	1205.5	0.6404	0.8308	1.4712
470	514.1	0.019 803	0.9025	450.9	668.4	1119.4	452.8	752.4	1205.2	0.6525	0.8093	1.4618
480	565.5	0.020 002	0.8187	462.2	656.7	1118.9	464.3	740.3	1204.6	0.6646	0.7878	1.4524
490	620.7	0.020 211	0.7436	473.6	644.7	1118.3	475.9	727.8	1203.7	0.6767	0.7663	1.4430

TABLE A-4E
(Continued)

Temp. °F T	Sat. press. psia P_{sat}	Specific volume ft³/lbm Sat. liquid v_f	Sat. vapor v_g	Internal energy Btu/lbm Sat. liquid u_f	Evap. u_{fg}	Sat. vapor u_g	Enthalpy Btu/lbm Sat. liquid h_f	Evap. h_{fg}	Sat. vapor h_g	Entropy Btu/(lbm·R) Sat. liquid s_f	Evap. s_{fg}	Sat. vapor s_g
500	680.0	0.020 43	0.6761	485.1	632.3	1117.4	487.7	714.8	1202.5	0.6888	0.7448	1.4335
520	811.4	0.020 91	0.5605	508.5	606.2	1114.8	511.7	687.3	1198.9	0.7130	0.7015	1.4145
540	961.5	0.021 45	0.4658	532.6	578.4	1111.0	536.4	657.5	1193.8	0.7374	0.6576	1.3950
560	1131.8	0.022 07	0.3877	557.4	548.4	1105.8	562.0	625.0	1187.0	0.7620	0.6129	1.3749
580	1324.3	0.022 78	0.3225	583.1	515.9	1098.9	588.6	589.3	1178.0	0.7872	0.5668	1.3540
600	1541.0	0.023 63	0.2677	609.9	480.1	1090.0	616.7	549.7	1166.4	0.8130	0.5187	1.3317
620	1784.4	0.024 65	0.2209	638.3	440.2	1078.5	646.4	505.0	1151.4	0.8398	0.4677	1.3075
640	2057.1	0.025 93	0.1805	668.7	394.5	1063.2	678.6	453.4	1131.9	0.8681	0.4122	1.2803
660	2362	0.027 67	0.144 59	702.3	340.0	1042.3	714.4	391.1	1105.5	0.8990	0.3493	1.2483
680	2705	0.030 32	0.111 27	741.7	269.3	1011.0	756.9	309.8	1066.7	0.9350	0.2718	1.2068
700	3090	0.036 66	0.074 38	801.7	145.9	947.7	822.7	167.5	990.2	0.9902	0.1444	1.1346
705.44	3204	0.050 53	0.050 53	872.6	0	872.6	902.5	0	902.5	1.0580	0	1.0580

Source: Tables A-4E through A-8E are adapted from Gordon J. Van Wylen and Richard E. Sonntag, *Fundamentals of Classical Thermodynamics,* English/SI Version, 3d ed., Wiley, New York, 1986, pp. 619-633. Originally published in Joseph H. Keenan, Frederick G. Keyes, Philip G. Hill, and Joan G. Moore, *Steam Tables,* Wiley, New York, 1969.

TABLE A-5E
Saturated water–Pressure table

Press. psia P	Sat. temp. °F T_{sat}	Specific volume ft³/lbm Sat. liquid v_f	Sat. vapor v_g	Internal energy Btu/lbm Sat. liquid u_f	Evap. u_{fg}	Sat. vapor u_g	Enthalpy Btu/lbm Sat. liquid h_f	Evap. h_{fg}	Sat. vapor h_g	Entropy Btu/(lbm·R) Sat. liquid s_f	Evap. s_{fg}	Sat. vapor s_g
1.0	101.70	0.016 136	333.6	69.74	974.3	1044.0	69.74	1036.0	1105.8	0.132 66	1.8453	1.9779
2.0	126.04	0.016 230	173.75	94.02	957.8	1051.8	94.02	1022.1	1116.1	0.174 99	1.7448	1.9198
3.0	141.43	0.016 300	118.72	109.38	947.2	1056.6	109.39	1013 1	1122.5	0.200 89	1.6852	1.8861
4.0	152.93	0.016 358	90.64	120.88	939.3	1060.2	120.89	1006.4	1127.3	0.219 83	1.6426	1.8624
5.0	162.21	0.016 407	73.53	130.15	932.9	1063.0	130.17	1000.9	1131.0	0.234 86	1.6093	1.8441
6.0	170.03	0.016 451	61.98	137.98	927.4	1065.4	138.00	996.2	1134.2	0.247 36	1.5819	1.8292
8.0	182.84	0.016 526	47.35	150.81	918.4	1069.2	150.84	988.4	1139.3	0.267 54	1.5383	1.8058
10	193.19	0.016 590	38.42	161.20	911.0	1072.2	161.23	982.1	1143.3	0.283 58	1.5041	1.7877
14.696	211.99	0.016 715	26.80	180.10	897.5	1077.6	180.15	970.4	1150.5	0.312 12	1.4446	1.7567
15	213.03	0.016 723	26.29	181.14	896.8	1077.9	181.19	969.7	1150.9	0.313 67	1.4414	1.7551
20	227.96	0.016 830	20.09	196.19	885.8	1.082.0	196.26	960.1	1156.4	0.335 80	1.3962	1.7320

Press. psia P	Sat. temp. °F T_{sat}	Specific volume ft³/lbm		Internal energy Btu/lbm			Enthalpy Btu/lbm			Entropy Btu/(lbm · R)		
		Sat. liquid v_f	Sat. vapor v_g	Sat. liquid u_f	Evap. u_{fg}	Sat. vapor u_g	Sat. liquid h_f	Evap. h_{fg}	Sat. vapor h_g	Sat. liquid s_f	Evap. s_{fg}	Sat. vapor s_g
25	240.08	0.016 922	16.306	208.44	876.9	1085.3	208.52	952.2	1160.7	0.353 45	1.3607	1.7142
30	250.34	0.017 004	13.748	218.84	869.2	1088.0	218.93	945.4	1164.3	0.368 21	1.3314	1.6996
35	259.30	0.017 073	11.900	227.93	862.4	1090.3	228.04	939.3	1167.4	0.380 93	1.3064	1.6873
40	267.26	0.017 146	10.501	236.03	856.2	1092.3	236.16	933.8	1170.0	0.392 14	1.2845	1.6767
45	274.46	0.017 209	9.403	243.37	850.7	1094.0	243.51	928.8	1172.3	0.402 18	1.2651	1.6673
50	281.03	0.017 269	8.518	250.08	845.5	1095.6	250.24	924.2	1174.4	0.411 29	1.2476	1.6589
55	287.10	0.017 325	7.789	256.28	840.8	1097.0	256.46	919.9	1176.3	0.419 63	1.2317	1.6513
60	292.73	0.017 378	7.177	262.06	836.3	1098.3	262.25	915.8	1178.0	0.427 33	1.2170	1.6444
65	298.00	0.017 429	6.657	267.46	832.1	1099.5	267.67	911.9	1179.6	0.434 50	1.2035	1.6380
70	302.96	0.017 478	6.209	272.56	828.1	1100.6	272.79	908.3	1181.0	0.441 20	1.1909	1.6321
75	307.63	0.017 524	5.818	277.37	824.3	1101.6	277.61	904.8	1182.4	0.447 49	1.1790	1.6265
80	312.07	0.017 570	5.474	281.95	820.6	1102.6	282.21	901.4	1183.6	0.453 44	1.1679	1.6214
85	316.29	0.017 613	5.170	286.30	817.1	1103.5	286.58	898.2	1184.8	0.459 07	1.1574	1.6165
90	320.31	0.017 655	4.898	290.46	813.8	1104.3	290.76	895.1	1185.9	0.464 42	1.1475	1.6119
95	324.16	0.017 696	4.654	294.45	810.6	1105.0	294.76	892.1	1186.9	0.469 52	1.1380	1.6076
100	327.86	0.017 736	4.434	298.28	807.5	1105.8	298.61	889.2	1187.8	0.474 39	1.1290	1.6034
110	334.82	0.017 813	4.051	305.52	801.6	1107.1	305.88	883.7	1189.6	0.483 55	1.1122	1.5957
120	341.30	0.017 886	3.730	312.27	796.0	1108.3	312.67	878.5	1191.1	0.492 01	1.0966	1.5886
130	347.37	0.017 957	3.457	318.61	790.7	1109.4	319.04	873.5	1192.5	0.499 89	1.0822	1.5821
140	353.08	0.018 024	3.221	324.58	785.7	1110.3	325.05	868.7	1193.8	0.507 27	1.0688	1.5761
150	358.48	0.018 089	3.016	330.24	781.0	1111.2	330.75	864.2	1194.9	0.514 22	1.0562	1.5704
160	363.60	0.018 152	2.836	335.63	776.4	1112.0	336.16	859.8	1196.0	0.520 78	1.0443	1.5651
170	368.47	0.018 214	2.676	340.76	772.0	1112.7	341.33	855.6	1196.9	0.527 00	1.0330	1.5600
180	373.13	0.018 273	2.533	345.68	767.7	1113.4	346.29	851.5	1197.8	0.532 92	1.0223	1.5553
190	377.59	0.018 331	2.405	350.39	763.6	1114.0	351.04	847.5	1198.6	0.538 57	1.0122	1.5507
200	381.86	0.018 387	2.289	354.9	759.6	1114.6	355.6	843.7	1199.3	0.5440	1.0025	1.5464
250	401.04	0.018 653	1.8448	375.4	741.4	1116.7	376.2	825.8	1202.1	0.5680	0.9594	1.5274
300	417.43	0.018 896	1.5442	393.0	725.1	1118.2	394.1	809.8	1203.9	0.5883	0.9232	1.5115
350	431.82	0.019 124	1.3267	408.7	710.3	1119.0	409.9	795.0	1204.9	0.6060	0.8917	1.4978
400	444.70	0.019 340	1.1620	422.8	696.7	1119.5	424.2	781.2	1205.5	0.6218	0.8638	1.4856
450	456.39	0.019 547	1.0326	435.7	683.9	1119.6	437.4	768.2	1205.6	0.6360	0.8385	1.4746
500	467.13	0.019 748	0.9283	447.7	671.7	1119.4	449.5	755.8	1205.3	0.6490	0.8154	1.4645
550	477.07	0.019 943	0.8423	458.9	660.2	1119.1	460.9	743.9	1204.8	0.6611	0.7941	1.4551
600	486.33	0.020 13	0.7702	469.4	649.1	1118.6	471.7	732.4	1204.1	0.6723	0.7742	1.4464
700	503.23	0.020 51	0.6558	488.9	628.2	1117.0	491.5	710.5	1202.0	0.6927	0.7378	1.4305
800	518.36	0.020 87	0.5691	506.6	608.4	1115.0	509.7	689.6	1199.3	0.7110	0.7050	1.4160
900	532.12	0.021 23	0.5009	523.0	589.6	1112.6	526.6	669.5	1196.0	0.7277	0.6750	1.4027
1000	544.75	0.021 59	0.4459	538.4	571.5	1109.9	542.4	650.0	1192.4	0.7432	0.6471	1.3903
1200	567.37	0.022 32	0.3623	566.7	536.8	1103.5	571.7	612.3	1183.9	0.7712	0.5961	1.3673
1400	587.25	0.023 07	0.3016	592.7	503.3	1096.0	598.6	575.5	1174.1	0.7964	0.5497	1.3461
1600	605.06	0.023 86	0.2552	616.9	470.5	1087.4	624.0	538.9	1162.9	0.8196	0.5062	1.3258
1800	621.21	0.024 72	0.2183	640.0	437.6	1077.7	648.3	502.1	1150.4	0.8414	0.4645	1.3060
2000	636.00	0.025 65	0.188 13	662.4	404.2	1066.6	671.9	464.4	1136.3	0.8623	0.4238	1.2861
2500	668.31	0.028 60	0.130 59	717.7	313.4	1031.0	730.9	360.5	1091.4	0.9131	0.3196	1.2327
3000	695.52	0.034 31	0.084 04	783.4	185.4	968.8	802.5	213.0	1015.5	0.9732	0.1843	1.1575
3203.6	705.44	0.050 53	0.050 53	872.6	0	872.6	902.5	0	902.5	1.0580	0	1.0580

TABLE A-6E
Superheated water

T °F	v ft³/lbm	u Btu/lbm	h Btu/lbm	s Btu/(lbm·R)	v ft³/lbm	u Btu/lbm	h Btu/lbm	s Btu/(lbm·R)	v ft³/lbm	u Btu/lbm	h Btu/lbm	s Btu/(lbm·R)
	P = 1.0 psia (101.70°F)*				*P* = 5.0 psia (162.21°F)				*P* = 10.0 psia (193.19°F)			
Sat.†	333.6	1044.0	1105.8	1.9779	73.53	1063.0	1131.0	1.8441	38.42	1072.2	1143.3	1.7877
200	392.5	1077.5	1150.1	2.0508	78.15	1076.3	1148.6	1.8715	38.85	1074.7	1146.6	1.7927
240	416.4	1091.2	1168.3	2.0775	83.00	1090.3	1167.1	1.8987	41.32	1089.0	1165.5	1.8205
280	440.3	1105.0	1186.5	2.1028	87.83	1104.3	1185.5	1.9244	43.77	1103.3	1184.3	1.8467
320	464.2	1118.9	1204.8	2.1269	92.64	1118.3	1204.0	1.9487	46.20	1117.6	1203.1	1.8714
360	488.1	1132.9	1223.2	2.1500	97.45	1132.4	1222.6	1.9719	48.62	1131.8	1221.8	1.8948
400	511.9	1147.0	1241.8	2.1720	102.24	1146.6	1241.2	1.9941	51.03	1146.1	1240.5	1.9171
440	535.8	1161.2	1260.4	2.1932	107.03	1160.9	1259.9	2.0154	53.44	1160.5	1259.3	1.9385
500	571.5	1182.8	1288.5	2.2235	114.20	1182.5	1288.2	2.0458	57.04	1182.2	1287.7	1.9690
600	631.1	1219.3	1336.1	2.2706	126.15	1219.1	1335.8	2.0930	63.03	1218.9	1335.5	2.0164
700	690.7	1256.7	1384.5	2.3142	138.08	1256.5	1384.3	2.1367	69.01	1256.3	1384.0	2.0601
800	750.3	1294.9	1433.7	2.3550	150.01	1294.7	1433.5	2.1775	74.98	1294.6	1433.3	2.1009
1000	869.5	1373.9	1534.8	2.4294	173.86	1373.9	1534.7	2.2520	86.91	1373.8	1534.6	2.1755
1200	988.6	1456.7	1639.6	2.4967	197.70	1456.6	1639.5	2.3192	98.84	1456.5	1639.4	2.2428
1400	1107.7	1543.1	1748.1	2.5584	221.54	1543.1	1748.1	2.3810	110.76	1543.0	1748.0	2.3045
	P = 14.696 psia (211.99°F)				*P* = 20 psia (227.96°F)				*P* = 40 psia (267.26°F)			
Sat.	26.80	1077.6	1150.5	1.7567	20.09	1082.0	1156.4	1.7320	10.501	1092.3	1170.0	1.6767
240	28.00	1087.9	1164.0	1.7764	20.47	1086.5	1162.3	1.7405				
280	29.69	1102.4	1183.1	1.8030	21.73	1101.4	1181.8	1.7676	10.711	1097.3	1176.6	1.6857
320	31.36	1116.8	1202.1	1.8280	22.98	1116.0	1201.0	1.7930	11.360	1112.8	1196.9	1.7124
360	33.02	1131.2	1221.0	1.8516	24.21	1130.6	1220.1	1.8168	11.996	1128.0	1216.8	1.7373
400	34.67	1145.6	1239.9	1.8741	25.43	1145.1	1239.2	1.8395	12.623	1143.0	1236.4	1.7606
440	36.31	1160.1	1258.8	1.8956	26.64	1159.6	1258.2	1.8611	13.243	1157.8	1255.8	1.7828
500	38.77	1181.8	1287.3	1.9263	28.46	1181.5	1286.8	1.8919	14.164	1180.1	1284.9	1.8140
600	42.86	1218.6	1335.2	1.9737	31.47	1218.4	1334.8	1.9395	15.685	1217.3	1333.4	1.8621
700	46.93	1256.1	1383.8	2.0175	34.47	1255.9	1383.5	1.9834	17.196	1255.1	1382.4	1.9063
800	51.00	1294.4	1433.1	2.0584	37.46	1294.3	1432.9	2.0243	18.701	1293.7	1432.1	1.9474
1000	59.13	1373.7	1534.5	2.1330	43.44	1373.5	1534.3	2.0989	21.70	1373.1	1533.8	2.0223
1200	67.25	1456.5	1639.3	2.2003	49.41	1456.4	1639.2	2.1663	24.69	1456.1	1638.9	2.0897
1400	75.36	1543.0	1747.9	2.2621	55.37	1542.9	1747.9	2.2281	27.68	1542.7	1747.6	2.1515
1600	83.47	1633.2	1860.2	2.3194	61.33	1633.2	1860.1	2.2854	30.66	1633.0	1859.9	2.2089
	P = 60 psia (292.73°F)				*P* = 80 psia (312.07°F)				*P* = 100 psia (327.86°F)			
Sat.	7.177	1098.3	1178.0	1.6444	5.474	1102.6	1183.6	1.6214	4.434	1105.8	1187.8	1.6034
320	7.485	1109.5	1192.6	1.6634	5.544	1106.0	1188.0	1.6271				
360	7.924	1125.3	1213.3	1.6893	5.886	1122.5	1209.7	1.6541	4.662	1119.7	1205.9	1.6259
400	8.353	1140.8	1233.5	1.7134	6.217	1138.5	1230.6	1.6790	4.934	1136.2	1227.5	1.6517
440	8.775	1156.0	1253.4	1.7360	6.541	1154.2	1251.0	1.7022	5.199	1152.3	1248.5	1.6755
500	9.399	1178.6	1283.0	1.7678	7.017	1177.2	1281.1	1.7346	5.587	1175.7	1279.1	1.7085
600	10.425	1216.3	1332.1	1.8165	7.794	1215.3	1330.7	1.7838	6.216	1214.2	1329.3	1.7582
700	11.440	1254.4	1381.4	1.8609	8.561	1253.6	1380.3	1.8285	6.834	1252.8	1379.2	1.8033
800	12.448	1293.0	1431.2	1.9022	9.321	1292.4	1430.4	1.8700	7.445	1291.8	1429.6	1.8449
1000	14.454	1372.7	1533.2	1.9773	10.831	1372.3	1532.6	1.9453	8.657	1371.9	1532.1	1.9204
1200	16.452	1455.8	1638.5	2.0448	12.333	1455.5	1638.1	2.0130	9.861	1455.2	1637.7	1.9882
1400	18.445	1542.5	1747.3	2.1067	13.830	1542.3	1747.0	2.0749	11.060	1542.0	1746.7	2.0502
1600	20.44	1632.8	1859.7	2.1641	15.324	1632.6	1859.5	2.1323	12.257	1632.4	1859.3	2.1076
1800	22.43	1726.7	1975.7	2.2179	16.818	1726.5	1975.5	2.1861	13.452	1726.4	1975.3	2.1614
2000	24.41	1824.0	2095.1	2.2685	18.310	1823.9	2094.9	2.2367	14.647	1823.7	2094.8	2.2121

*The temperature in parentheses is the saturation temperature at the specified pressure.

†Properties of saturated vapor at the specified pressure.

T °F	v ft³/lbm	u Btu/lbm	h Btu/lbm	s Btu/(lbm·R)	v ft³/lbm	u Btu/lbm	h Btu/lbm	s Btu/(lbm·R)	v ft³/lbm	u Btu/lbm	h Btu/lbm	s Btu/(lbm·R)
	P = 120 psia (341.30°F)				*P* = 140 psia (353.08°F)				*P* = 160 psia (363.60°F)			
Sat.	3.730	1108.3	1191.1	1.5886	3.221	1110.3	1193.8	1.5761	2.836	1112.0	1196.0	1.5651
360	3.844	1116.7	1202.0	1.6021	3.259	1113.5	1198.0	1.5812				
400	4.079	1133.8	1224.4	1.6288	3.466	1131.4	1221.2	1.6088	3.007	1128.8	1217.8	1.5911
450	4.360	1154.3	1251.2	1.6590	3.713	1152.4	1248.6	1.6399	3.228	1150.5	1246.1	1.6230
500	4.633	1174.2	1277.1	1.6868	3.952	1172.7	1275.1	1.6682	3.440	1171.2	1273.0	1.6518
550	4.900	1193.8	1302.6	1.7127	4.184	1192.6	1300.9	1.6944	3.646	1191.3	1299.2	1.6784
600	5.164	1213.2	1327.8	1.7371	4.412	1212.1	1326.4	1.7191	3.848	1211.1	1325.0	1.7034
700	5.682	1252.0	1378.2	1.7825	4.860	1251.2	1377.1	1.7648	4.243	1250.4	1376.0	1.7494
800	6.195	1291.2	1428.7	1.8243	5.301	1290.5	1427.9	1.8068	4.631	1289.9	1427.0	1.7916
1000	7.208	1371.5	1531.5	1.9000	6.173	1371.0	1531.0	1.8827	5.397	1370.6	1530.4	1.8677
1200	8.213	1454.9	1637.3	1.9679	7.036	1454.6	1636.9	1.9507	6.154	1454.3	1636.5	1.9358
1400	9.214	1541.8	1746.4	2.0300	7.895	1541.6	1746.1	2.0129	6.906	1541.4	1745.9	1.9980
1600	10.212	1632.3	1859.0	2.0875	8.752	1632.1	1858.8	2.0704	7.656	1631.9	1858.6	2.0556
1800	11.209	1726.2	1975.1	2.1413	9.607	1726.1	1975.0	2.1242	8.405	1725.9	1974.8	2.1094
2000	12.205	1823.6	2094.6	2.1919	10.461	1823.5	2094.5	2.1749	9.153	1823.3	2094.3	2.1601
	P = 180 psia (373.13°F)				*P* = 200 psia (381.86°F)				*P* = 225 psia (391.87°F)			
Sat.	2.533	1113.4	1197.8	1.5553	2.289	1114.6	1199.3	1.5464	2.043	1115.8	1200.8	1.5365
400	2.648	1126.2	1214.4	1.5749	2.361	1123.5	1210.8	1.5600	2.073	1119.9	1206.2	1.5427
450	2.850	1148.5	1243.4	1.6078	2.548	1146.4	1240.7	1.5938	2.245	1143.8	1237.3	1.5779
500	3.042	1169.6	1270.9	1.6372	2.724	1168.0	1268.8	1.6239	2.405	1165.9	1266.1	1.6087
550	3.228	1190.0	1297.5	1.6642	2.893	1188.7	1295.7	1.6512	2.588	1187.0	1293.5	1.6366
600	3.409	1210.0	1323.5	1.6893	3.058	1208.9	1322.1	1.6767	2.707	1207.5	1320.2	1.6624
700	3.763	1249.6	1374.9	1.7357	3.379	1248.8	1373.8	1.7234	2.995	1247.7	1372.4	1.7095
800	4.110	1289.3	1426.2	1.7781	3.693	1288.6	1425.3	1.7660	3.276	1287.8	1424.2	1.7523
900	4.453	1329.4	1477.7	1.8175	4.003	1328.9	1477.1	1.8055	3.553	1328.3	1476.2	1.7920
1000	4.793	1370.2	1529.8	1.8545	4.310	1369.8	1529.3	1.8425	3.827	1369.3	1528.6	1.8292
1200	5.467	1454.0	1636.1	1.9227	4.918	1453.7	1635.7	1.9109	4.369	1453.4	1635.3	1.8977
1400	6.137	1541.2	1745.6	1.9849	5.521	1540.9	1745.3	1.9732	4.906	1540.7	1744.9	1.9600
1600	6.804	1631.7	1858.4	2.0425	6.123	1631.6	1858.2	2.0308	5.441	1631.3	1857.9	2.0177
1800	7.470	1725.8	1974.6	2.0964	6.722	1725.6	1974.4	2.0847	5.975	1725.4	1974.2	2.0716
2000	8.135	1823.2	2094.2	2.1470	7.321	1823.0	2094.0	2.1354	6.507	1822.9	2093.8	2.1223
	P = 250 psia (401.04°F)				*P* = 275 psia (409.52°F)				*P* = 300 psia (417.43°F)			
Sat.	1.8448	1116.7	1202.1	1.5274	1.6813	1117.5	1203.1	1.5192	1.5442	1118.2	1203.9	1.5115
450	2.002	1141.1	1233.7	1.5632	1.8026	1138.3	1230.0	1.5495	1.6361	1135.4	1226.2	1.5365
500	2.150	1163.8	1263.3	1.5948	1.9407	1161.7	1260.4	1.5820	1.7662	1159.5	1257.5	1.5701
550	2.290	1185.3	1291.3	1.6233	2.071	1183.6	1289.0	1.6110	1.8878	1181.9	1286.7	1.5997
600	2.426	1206.1	1318.3	1.6494	2.196	1204.7	1316.4	1.6376	2.004	1203.2	1314.5	1.6266
650	2.558	1226.5	1344.9	1.6739	2.317	1225.3	1343.2	1.6623	2.117	1224.1	1341.6	1.6516
700	2.688	1246.7	1371.1	1.6970	2.436	1245.7	1369.7	1.6856	2.227	1244.6	1368.3	1.6751
800	2.943	1287.0	1423.2	1.7401	2.670	1286.2	1422.1	1.7289	2.442	1285.4	1421.0	1.7187
900	3.193	1327.6	1475.3	1.7799	2.898	1327.0	1474.5	1.7689	2.653	1326.3	1473.6	1.7589
1000	3.440	1368.7	1527.9	1.8172	3.124	1368.2	1527.2	1.8064	2.860	1367.7	1526.5	1.7964
1200	3.929	1453.0	1634.8	1.8858	3.570	1452.6	1634.3	1.8751	3.270	1452.2	1633.8	1.8653
1400	4.414	1540.4	1744.6	1.9483	4.011	1540.1	1744.2	1.9376	3.675	1539.8	1743.8	1.9279
1600	4.896	1631.1	1857.6	2.0060	4.450	1630.9	1857.3	1.9954	4.078	1630.7	1857.0	1.9857
1800	5.376	1725.2	1974.0	2.0599	4.887	1725.0	1973.7	2.0493	4.479	1724.9	1973.5	2.0396
2000	5.856	1822.7	2093.6	2.1106	5.323	1822.5	2093.4	2.1000	4.879	1822.3	2093.2	2.0904

TABLE A-6E
(*Continued*)

T °F	v ft³/lbm	u Btu/lbm	h Btu/lbm	s Btu/(lbm·R)	v ft³/lbm	u Btu/lbm	h Btu/lbm	s Btu/(lbm·R)	v ft³/lbm	u Btu/lbm	h Btu/lbm	s Btu/(lbm·R)
	P = 350 psia (431.82°F)				P = 400 psia (444.70°F)				P = 450 psia (456.39°F)			
Sat.	1.3267	1119.0	1204.9	1.4978	1.1620	1119.5	1205.5	1.4856	1.0326	1119.6	1205.6	1.4746
450	1.3733	1129.2	1218.2	1.5125	1.1745	1122.6	1209.6	1.4901				
500	1.4913	1154.9	1251.5	1.5482	1.2843	1150.1	1245.2	1.5282	1.1226	1145.1	1238.5	1.5097
550	1.5998	1178.3	1281.9	1.5790	1.3833	1174.6	1277.0	1.5605	1.2146	1170.7	1271.9	1.5436
600	1.7025	1200.3	1310.6	1.6068	1.4760	1197.3	1306.6	1.5892	1.2996	1194.3	1302.5	1.5732
650	1.8013	1221.6	1338.3	1.6323	1.5645	1219.1	1334.9	1.6153	1.3803	1216.6	1331.5	1.6000
700	1.8975	1242.5	1365.4	1.6562	1.6503	1240.4	1362.5	1.6397	1.4580	1238.2	1359.6	1.6248
800	2.085	1283.8	1418.8	1.7004	1.8163	1282.1	1416.6	1.6844	1.6077	1280.5	1414.4	1.6701
900	2.267	1325.0	1471.8	1.7409	1.9776	1323.7	1470.1	1.7252	1.7524	1322.4	1468.3	1.7113
1000	2.446	1366.6	1525.0	1.7787	2.136	1365.5	1523.6	1.7632	1.8941	1364.4	1522.2	1.7495
1200	2.799	1451.5	1632.8	1.8478	2.446	1450.7	1631.8	1.8327	2.172	1450.0	1630.8	1.8192
1400	3.148	1539.3	1743.1	1.9106	2.752	1538.7	1742.4	1.8956	2.444	1538.1	1741.7	1.8823
1600	3.494	1630.2	1856.5	1.9685	3.055	1629.8	1855.9	1.9535	2.715	1629.3	1855.4	1.9403
1800	3.838	1724.5	1973.1	2.0225	3.357	1724.1	1972.6	2.0076	2.983	1723.7	1972.1	1.9944
2000	4.182	1822.0	2092.8	2.0733	3.658	1821.6	2092.4	2.0584	3.251	1821.3	2092.0	2.0453
	P = 500 psia (467.13°F)				P = 600 psia (486.33°F)				P = 700 psia (503.23°F)			
Sat.	0.9283	1119.4	1205.3	1.4645	0.7702	1118.6	1204.1	1.4464	0.6558	1117.0	1202.0	1.4305
500	0.9924	1139.7	1231.5	1.4923	0.7947	1128.0	1216.2	1.4592				
550	1.0792	1166.7	1266.6	1.5279	0.8749	1158.2	1255.4	1.4990	0.7275	1149.0	1243.2	1.4723
600	1.1583	1191.1	1298.3	1.5585	0.9456	1184.5	1289.5	1.5320	0.7929	1177.5	1280.2	1.5081
650	1.2327	1214.0	1328.0	1.5860	1.0109	1208.6	1320.9	1.5609	0.8520	1203.1	1313.4	1.5387
700	1.3040	1236.0	1356.7	1.6112	1.0727	1231.5	1350.6	1.5872	0.9073	1226.9	1344.4	1.5661
800	1.4407	1278.8	1412.1	1.6571	1.1900	1275.4	1407.6	1.6343	1.0109	1272.0	1402.9	1.6145
900	1.5723	1321.0	1466.5	1.6987	1.3021	1318.4	1462.9	1.6766	1.1089	1315.6	1459.3	1.6576
1000	1.7008	1363.3	1520.7	1.7371	1.4108	1361.2	1517.8	1.7155	1.2036	1358.9	1514.9	1.6970
1100	1.8271	1406.0	1575.1	1.7731	1.5173	1404.2	1572.7	1.7519	1.2960	1402.4	1570.2	1.7337
1200	1.9518	1449.2	1629.8	1.8072	1.6222	1447.7	1627.8	1.7861	1.3868	1446.2	1625.8	1.7682
1400	2.198	1537.6	1741.0	1.8704	1.8289	1536.5	1739.5	1.8497	1.5652	1535.3	1738.1	1.8321
1600	2.442	1628.9	1854.8	1.9285	2.033	1628.0	1853.7	1.9080	1.7409	1627.1	1852.6	1.8906
1800	2.684	1723.3	1971.7	1.9827	2.236	1722.6	1970.8	1.9622	1.9152	1721.8	1969.9	1.9449
2000	2.926	1820.9	2091.6	2.0335	2.438	1820.2	2090.8	2.0131	2.0887	1819.5	2090.1	1.9958
	P = 800 psia (518.36°F)				P = 1000 psia (544.75°F)				P = 1250 psia (572.56°F)			
Sat.	0.5691	1115.0	1199.3	1.4160	0.4459	1109.9	1192.4	1.3903	0.3454	1101.7	1181.6	1.3619
550	0.6154	1138.8	1229.9	1.4469	0.4534	1114.8	1198.7	1.3966				
600	0.6776	1170.1	1270.4	1.4861	0.5140	1153.7	1248.8	1.4450	0.3786	1129.0	1216.6	1.3954
650	0.7324	1197.2	1305.6	1.5186	0.5637	1184.7	1289.1	1.4822	0.4267	1167.2	1266.0	1.4410
700	0.7829	1222.1	1338.0	1.5471	0.6080	1212.0	1324.6	1.5135	0.4670	1198.4	1306.4	1.4767
750	0.8306	1245.7	1368.6	1.5730	0.6490	1237.2	1357.3	1.5412	0.5030	1226.1	1342.4	1.5070
800	0.8764	1268.5	1398.2	1.5969	0.6878	1261.2	1388.5	1.5664	0.5364	1251.8	1375.8	1.5341
900	0.9640	1312.9	1455.6	1.6408	0.7610	1307.3	1448.1	1.6120	0.5984	1300.0	1438.4	1.5820
1000	1.0482	1356.7	1511.9	1.6807	0.8305	1352.2	1505.9	1.6530	0.6563	1346.4	1498.2	1.6244
1100	1.1300	1400.5	1567.8	1.7178	0.8976	1396.8	1562.9	1.6908	0.7116	1392.0	1556.6	1.6631
1200	1.2102	1444.6	1623.8	1.7526	0.9630	1441.5	1619.7	1.7261	0.7652	1437.5	1614.5	1.6991
1400	1.3674	1534.2	1736.6	1.8167	1.0905	1531.9	1733.7	1.7909	0.8689	1529.0	1730.0	1.7648
1600	1.5218	1626.2	1851.5	1.8754	1.2152	1624.4	1849.3	1.8499	0.9699	1622.2	1846.5	1.8243
1800	1.6749	1721.0	1969.0	1.9298	1.3384	1719.5	1967.2	1.9046	1.0693	1717.6	1965.0	1.8791
2000	1.8271	1818.8	2089.3	1.9808	1.4608	1817.4	2087.7	1.9557	1.1678	1815.7	2085.8	1.9304

938

T °F	v ft³/ lbm	u Btu/ lbm	h Btu/ lbm	s Btu/ (lbm · R)	v ft³/ lbm	u Btu/ lbm	h Btu/ lbm	s Btu/ (lbm · R)	v ft³/ lbm	u Btu/ lbm	h Btu/ lbm	s Btu/ (lbm · R)
	P = 1500 psia (596.39°F)				*P* = 1750 psia (617.31°F)				*P* = 2000 psia (636.00°F)			
Sat.	0.2769	1091.8	1168.7	1.3359	0.2268	1080.2	1153.7	1.3109	0.188 13	1066.6	1136.3	1.2861
600	0.2816	1096.6	1174.8	1.3416								
650	0.3329	1147.0	1239.4	1.4012	0.2627	1122.5	1207.6	1.3603	0.2057	1091.1	1167.2	1.3141
700	0.3716	1183.4	1286.6	1.4429	0.3022	1166.7	1264.6	1.4106	0.2487	1147.7	1239.8	1.3782
750	0.4049	1214.1	1326.5	1.4767	0.3341	1201.3	1309.5	1.4485	0.2803	1187.3	1291.1	1.4216
800	0.4350	1241.8	1362.5	1.5058	0.3622	1231.3	1348.6	1.4802	0.3071	1220.1	1333.8	1.4562
850	0.4631	1267.7	1396.2	1.5320	0.3878	1258.8	1384.4	1.5081	0.3312	1249.5	1372.0	1.4860
900	0.4897	1292.5	1428.5	1.5562	0.4119	1284.8	1418.2	1.5334	0.3534	1276.8	1407.6	1.5126
1000	0.5400	1340.4	1490.3	1.6001	0.4569	1334.3	1482.3	1.5789	0.3945	1328.1	1474.1	1.5598
1100	0.5876	1387.2	1550.3	1.6399	0.4990	1382.2	1543.8	1.6197	0.4325	1377.2	1537.2	1.6017
1200	0.6334	1433.5	1609.3	1.6765	0.5392	1429.4	1604.0	1.6571	0.4685	1425.2	1598.6	1.6398
1400	0.7213	1526.1	1726.3	1.7431	0.6158	1523.1	1722.6	1.7245	0.5368	1520.2	1718.8	1.7082
1600	0.8064	1619.9	1843.7	1.8031	0.6896	1617.6	1841.0	1.7850	0.6020	1615.4	1838.2	1.7692
1800	0.8899	1715.7	1962.7	1.8582	0.7617	1713.9	1960.5	1.8404	0.6656	1712.0	1958.3	1.8249
2000	0.9725	1814.0	2083.9	1.9096	0.8330	1812.3	2082.0	1.8919	0.7284	1810.6	2080.2	1.8765
	P = 2500 psia (668.31°F)				*P* = 3000 psia (695.52°F)				*P* = 3500 psia			
Sat.	0.130 59	1031.0	1091.4	1.2327	0.084 04	968.8	1015.5	1.1575				
650									0.024 91	663.5	679.7	0.8630
700	0.168 39	1098.7	1176.6	1.3073	0.097 71	1003.9	1058.1	1.1944	0.030 58	759.5	779.3	0.9506
750	0.2030	1155.2	1249.1	1.3686	0.148 31	1114.7	1197.1	1.3122	0.104 60	1058.4	1126.1	1.2440
800	0.2291	1195.7	1301.7	1.4112	0.175 72	1167.6	1265.2	1.3675	0.136 26	1134.7	1223.0	1.3226
850	0.2513	1229.5	1345.8	1.4456	0.197 31	1207.7	1317.2	1.4080	0.158 18	1183.4	1285.9	1.3716
900	0.2712	1259.5	1385.4	1.4752	0.2160	1241.8	1361.7	1.4414	0.176 25	1222.4	1336.5	1.4096
950	0.2896	1288.2	1422.2	1.5018	0.2328	1272.7	1402.0	1.4705	0.192 14	1256.4	1380.8	1.4416
1000	0.3069	1315.2	1457.2	1.5262	0.2485	1301.7	1439.6	1.4967	0.2066	1287.6	1421.4	1.4699
1100	0.3393	1366.8	1523.8	1.5704	0.2772	1356.2	1510.1	1.5434	0.2328	1345.2	1496.0	1.5193
1200	0.3696	1416.7	1587.7	1.6101	0.3036	1408.0	1576.6	1.5848	0.2566	1399.2	1565.3	1.5624
1400	0.4261	1514.2	1711.3	1.6804	0.3524	1508.1	1703.7	1.6571	0.2997	1501.9	1696.1	1.6368
1600	0.4795	1610.2	1832.6	1.7424	0.3978	1606.3	1827.1	1.7201	0.3395	1601.7	1821.6	1.7010
1800	0.5312	1708.2	1954.0	1.7986	0.4416	1704.5	1949.6	1.7769	0.3776	1700.8	1945.4	1.7583
2000	0.5820	1807.2	2076.4	1.8506	0.4844	1803.9	2072.8	1.8291	0.4147	1800.6	2069.2	1.8108
	P = 4000 psia				*P* = 5000 psia				*P* = 6000 psia			
650	0.024 47	657.7	675.8	0.8574	0.023 77	648.0	670.0	0.8482	0.012 22	640.0	665.8	0.8405
700	0.028 67	742.1	763.4	0.9345	0.026 76	721.8	746.6	0.9156	0.025 63	708.1	736.5	0.9028
750	0.063 31	960.7	1007.5	1.1395	0.033 64	821.4	852.6	1.0049	0.029 78	788.6	821.7	0.9746
800	0.105 22	1095.0	1172.9	1.2740	0.059 32	987.2	1042.1	1.1583	0.039 42	896.9	940.7	1.0708
850	0.128 33	1156.5	1251.5	1.3352	0.085 56	1092.7	1171.9	1.2596	0.058 18	1018.8	1083.4	1.1820
900	0.146 22	1201.5	1309.7	1.3789	0.103 85	1155.1	1251.1	1.3190	0.075 88	1102.9	1187.2	1.2599
950	0.161 51	1239.2	1358.8	1.4144	0.118 53	1202.2	1311.9	1.3629	0.090 08	1162.0	1262.0	1.3140
1000	0.175 20	1272.9	1402.6	1.4449	0.131 20	1242.0	1363.4	1.3988	0.102 07	1209.1	1322.4	1.3561
1100	0.199 54	1333.9	1481.6	1.4973	0.153 02	1310.6	1452.2	1.4577	0.122 18	1286.4	1422.1	1.4222
1200	0.2213	1390.1	1553.9	1.5423	0.171 99	1371.6	1530.8	1.5066	0.139 27	1352.7	1507.3	1.4752
1300	0.2414	1443.7	1622.4	1.5823	0.189 18	1428.6	1603.7	1.5493	0.154 53	1413.3	1584.9	1.5206
1400	0.2603	1495.7	1688.4	1.6188	0.205 17	1483.2	1673.0	1.5876	0.168 54	1470.5	1657.6	1.5608
1600	0.2959	1597.1	1816.1	1.6841	0.2348	1587.9	1805.2	1.6551	0.194 20	1578.7	1794.3	1.6307
1800	0.3296	1697.1	1941.1	1.7420	0.2626	1689.8	1932.7	1.7142	0.218 01	1682.4	1924.5	1.6910
2000	0.3625	1797.3	2065.6	1.7948	0.2895	1790.8	2058.6	1.7676	0.240 87	1784.3	2051.7	1.7450

TABLE A-7E
Compressed liquid water

T °F	v ft³/lbm	u Btu/lbm	h Btu/lbm	s Btu/(lbm·R)	v ft³/lbm	u Btu/lbm	h Btu/lbm	s Btu/(lbm·R)	v ft³/lbm	u Btu/lbm	h Btu/lbm	s Btu/(lbm·R)
	P = 500 psia (467.13°F)				P = 1000 psia (544.75°F)				P = 1500 psia (596.39°F)			
Sat.	0.019 748	447.70	449.53	0.649 04	0.021 591	538.39	542.38	0.743 20	0.023 461	604.97	611.48	0.808 24
32	0.015 994	0.00	1.49	0.000 00	0.015 967	0.03	2.99	0.000 05	0.015 939	0.05	4.47	0.000 07
50	0.015 998	18.02	19.50	0.035 99	0.015 972	17.99	20.94	0.035 92	0.015 946	17.95	22.38	0.035 84
100	0.016 106	67.87	69.36	0.129 32	0.016 082	67.70	70.68	0.129 01	0.016 058	67.53	71.99	0.128 70
150	0.016 318	117.66	119.17	0.214 57	0.016 293	117.38	120.40	0.214 10	0.016 268	117.10	121.62	0.213 64
200	0.016 608	167.65	169.19	0.293 41	0.016 580	167.26	170.32	0.292 81	0.016 554	166.87	171.46	0.292 21
250	0.016 972	217.99	219.56	0.367 02	0.016 941	217.47	220.61	0.366 28	0.016 910	216.96	221.65	0.365 54
300	0.017 416	268.92	270.53	0.436 41	0.017 379	268.24	271.46	0.435 52	0.017 343	267.58	272.39	0.434 63
350	0.017 954	320.71	322.37	0.502 49	0.017 909	319.83	323.15	0.501 40	0.017 865	318.98	323.94	0.500 34
400	0.018 608	373.68	375.40	0.566 04	0.018 550	372.55	375.98	0.564 72	0.018 493	371.45	376.59	0.563 43
450	0.019 420	428.40	430.19	0.627 98	0.019 340	426.89	430.47	0.626 32	0.019 264	425.44	430.79	0.624 70
500					0.020 36	483.8	487.5	0.6874	0.020 24	481.8	487.4	0.6853
550									0.021 58	542.1	548.1	0.7469
	P = 2000 psia (636.00°F)				P = 3000 psia (695.52°F)				P = 5000 psia			
Sat.	0.025 649	662.40	671.89	0.862 27	0.034 310	783.45	802.50	0.973 20				
32	0.015 912	0.06	5.95	0.000 08	0.015 859	0.09	8.90	0.000 09	0.015 755	0.11	14.70	−0.000 01
50	0.015 920	17.91	23.81	0.035 75	0.015 870	17.84	26.65	0.035 55	0.015 773	17.67	32.26	0.035 08
100	0.016 034	67.37	73.30	0.128 39	0.015 987	67.04	75.91	0.127 77	0.015 897	66.40	81.11	0.126 51
200	0.016 527	166.49	172.60	0.291 62	0.016 476	165.74	174.89	0.290 46	0.016 376	164.32	179.47	0.288 18
300	0.017 308	266.93	273.33	0.433 76	0.017 240	265.66	275.23	0.432 05	0.017 110	263.25	279.08	0.428 75
400	0.018 439	370.38	377.21	0.562 16	0.018 334	368.32	378.50	0.559 70	0.018 141	364.47	381.25	0.555 06
450	0.019 191	424.04	431.14	0.623 13	0.019 053	421.36	431.93	0.620 11	0.018 803	416.44	433.84	0.614 51
500	0.020 14	479.8	487.3	0.6832	0.019 944	476.2	487.3	0.6794	0.019 603	469.8	487.9	0.6724
560	0.021 72	551.8	559.8	0.7565	0.021 382	546.2	558.0	0.7508	0.020 835	536.7	556.0	0.7411
600	0.023 30	605.4	614.0	0.8086	0.022 74	597.0	609.6	0.8004	0.021 91	584.0	604.2	0.7876
640					0.024 75	654.3	668.0	0.8545	0.023 34	634.6	656.2	0.8357
680					0.028 79	728.4	744.3	0.9226	0.025 35	690.6	714.1	0.8873
700									0.026 76	721.8	746.6	0.9156

Temp. °F T	Sat. Press. psia P_{sat}	Specific volume ft³/lbm		Internal energy Btu/lbm			Enthalpy Btu/lbm			Entropy Btu/(lbm · R)		
		Sat. ice v_i	Sat. vapor $v_g \times 10^{-3}$	Sat. ice u_i	Subl. u_{ig}	Sat. vapor u_g	Sat. ice h_i	Subl. h_{ig}	Sat. vapor h_g	Sat. ice s_i	Subl. s_{ig}	Sat. vapor s_g
32.018	0.0887	0.017 47	3.302	−143.34	1164.6	1021.2	−143.34	1218.7	1075.4	−0.292	2.479	2.187
32	0.0886	0.017 47	3.305	−143.35	1164.6	1021.2	−143.35	1218.7	1075.4	−0.292	2.479	2.187
30	0.0808	0.017 47	3.607	−144.35	1164.9	1020.5	−144.35	1218.9	1074.5	−0.294	2.489	2.195
25	0.0641	0.017 46	4.506	−146.84	1165.7	1018.9	−146.84	1219.1	1072.3	−0.299	2.515	2.216
20	0.0505	0.017 45	5.655	−149.31	1166.5	1017.2	−149.31	1219.4	1070.1	−0.304	2.542	2.238
15	0.0396	0.017 45	7.13	−151.75	1167.3	1015.5	−151.75	1219.7	1067.9	−0.309	2.569	2.260
10	0.0309	0.017 44	9.04	−154.17	1168.1	1013.9	−154.17	1219.9	1065.7	−0.314	2.597	2.283
5	0.0240	0.017 43	11.52	−156.56	1168.8	1012.2	−156.56	1220.1	1063.5	−0.320	2.626	2.306
0	0.0185	0.017 43	14.77	−158.93	1169.5	1010.6	−158.93	1220.2	1061.2	−0.325	2.655	2.330
−5	0.0142	0.017 42	19.03	−161.27	1170.2	1008.9	−161.27	1220.3	1059.0	−0.330	2.684	2.354
−10	0.0109	0.017 41	24.66	−163.59	1170.9	1007.3	−163.59	1220.4	1056.8	−0.335	2.714	2.379
−15	0.0082	0.017 40	32.2	−165.89	1171.5	1005.6	−165.89	1220.5	1054.6	−0.340	2.745	2.405
−20	0.0062	0.017 40	42.2	−168.16	1172.1	1003.9	−168.16	1220.6	1052.4	−0.345	2.776	2.431
−25	0.0046	0.017 39	55.7	−170.40	1172.7	1002.3	−170.40	1220.6	1050.2	−0.351	2.808	2.457
−30	0.0035	0.017 38	74.1	−172.63	1173.2	1000.6	−172.63	1220.6	1048.0	−0.356	2.841	2.485
−35	0.0026	0.017 37	99.2	−174.82	1173.8	988.9	−174.82	1220.6	1045.8	−0.361	2.874	2.513
−40	0.0019	0.017 37	133.8	−177.00	1174.3	997.3	−177.00	1220.6	1043.6	−0.366	2.908	2.542

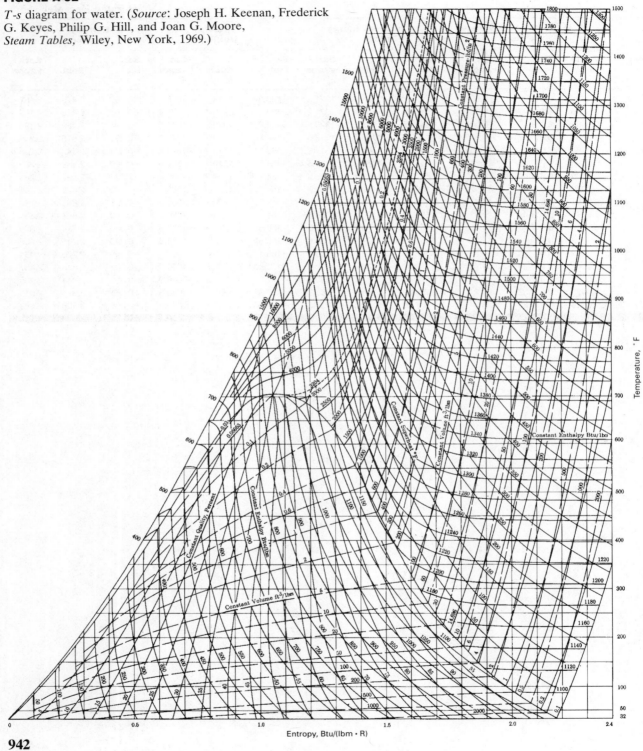

T-s diagram for water. (*Source*: Joseph H. Keenan, Frederick G. Keyes, Philip G. Hill, and Joan G. Moore, *Steam Tables,* Wiley, New York, 1969.)

Entropy, Btu/(lbm · R)

Temperature, °F

Mollier diagram for water. (*Source*: Joseph H. Keenan, Frederick G. Keyes, Philip G. Hill, and Joan G. Moore, *Steam Tables,* Wiley, New York, 1969.)

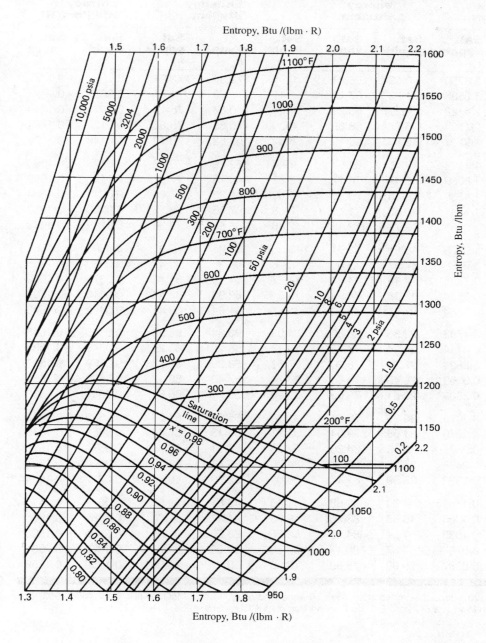

Entropy, Btu /(lbm · R)

943

TABLE A-11E
Saturated refrigerant-12–Temperature table

Temp. °F T	Sat. press. psia P_{sat}	Specific volume ft³/lbm		Internal energy Btu/lbm		Enthalpy Btu/lbm			Entropy Btu/(lbm·R)	
		Sat. liquid v_f	Sat. vapor v_g	Sat. liquid u_f	Sat. vapor u_g	Sat. liquid h_f	Evap. h_{fg}	Sat. vapor h_g	Sat. liquid s_f	Sat. vapor s_g
−40	9.308	0.010 56	3.8750	−0.02	66.24	0	72.91	72.91	0	0.1737
−30	11.999	0.010 67	3.0585	1.93	67.22	2.11	71.90	74.01	0.0050	0.1723
−20	15.267	0.010 79	2.4429	4.21	68.21	4.24	70.87	75.11	0.0098	0.1710
−15	17.141	0.010 85	2.1924	5.27	68.69	5.30	70.35	75.65	0.0122	0.1704
−10	19.189	0.010 91	1.9727	6.33	69.19	6.37	69.82	76.19	0.0146	0.1699
−5	21.422	0.010 97	1.7794	7.40	69.67	7.44	69.29	76.73	0.0170	0.1694
0	23.849	0.011 03	1.6089	8.47	70.17	8.52	68.75	77.27	0.0193	0.1689
5	26.483	0.011 09	1.4580	9.55	70.65	9.60	68.20	77.80	0.0216	0.1684
10	29.335	0.011 16	1.3241	10.62	71.15	10.68	67.65	78.33	0.0240	0.1680
20	35.736	0.011 30	1.0988	12.79	72.12	12.86	66.52	79.38	0.0285	0.1672
25	39.310	0.011 37	1.0039	13.88	72.61	13.96	65.95	79.91	0.0308	0.1668
30	43.148	0.011 44	0.9188	14.97	73.08	15.06	65.36	80.42	0.0330	0.1665
40	51.667	0.011 59	0.7736	17.16	74.04	17.27	64.16	81.43	0.0375	0.1659
50	61.394	0.011 75	0.6554	19.38	74.99	19.51	62.93	82.44	0.0418	0.1653
60	72.433	0.011 91	0.5584	21.61	75.92	21.77	61.64	83.41	0.0462	0.1648
70	84.89	0.012 09	0.4782	23.86	76.85	24.05	60.31	84.36	0.0505	0.1643
80	98.87	0.012 28	0.4114	26.14	77.76	26.37	58.92	85.29	0.0548	0.1639
85	106.47	0.012 38	0.3821	27.29	78.20	27.53	58.20	85.73	0.0569	0.1637
90	114.49	0.012 48	0.3553	28.45	78.65	28.71	57.46	86.17	0.0590	0.1635
95	122.95	0.012 58	0.3306	29.61	79.09	29.90	56.71	86.61	0.0611	0.1633
100	131.86	0.012 69	0.3079	30.79	79.51	31.10	55.93	87.03	0.0632	0.1632
105	141.25	0.012 81	0.2870	31.98	79.94	32.31	55.13	87.44	0.0653	0.1630
110	151.11	0.012 92	0.2677	33.16	80.36	33.53	54.31	87.84	0.0675	0.1628
115	161.47	0.013 05	0.2498	34.37	80.76	34.76	53.47	88.23	0.0696	0.1626
120	172.35	0.013 17	0.2333	35.59	81.17	36.01	52.60	88.61	0.0717	0.1624
140	221.32	0.013 75	0.1780	40.60	82.68	41.16	48.81	89.97	0.0802	0.1616
160	279.82	0.014 45	0.1360	45.88	83.96	46.63	44.37	91.00	0.0889	0.1605
180	349.00	0.015 36	0.1033	51.57	84.89	52.56	39.00	91.56	0.0980	0.1590
200	430.09	0.016 66	0.0767	57.87	85.17	59.20	32.08	91.28	0.1079	0.1565
233.6	596.9	0.0287	0.0287	75.69	75.69	78.86	0	78.86	0.1359	0.1359

Source: Tables A-11E through A-13E are adapted from Kenneth Wark, *Thermodynamics,* 4th ed., McGraw-Hill, New York, 1983, pp. 857–861. Originally published by Freon Products Division, E. I. du Pont de Nemours & Company, 1956.

Press. psia P	Sat. Temp. °F T_{sat}	Specific volume ft³/lbm		Internal energy Btu/lbm		Enthalpy Btu/lbm			Entropy Btu/(lbm·R)	
		Sat. liquid v_f	Sat. vapor v_g	Sat. liquid u_f	Sat. vapor u_g	Sat. liquid h_f	Evap. h_{fg}	Sat. vapor h_g	Sat. liquid s_f	Sat. vapor s_g
5	−62.35	0.0103	6.9069	−4.69	64.04	−4.68	75.11	70.43	−0.0114	0.1776
10	−37.23	0.0106	3.6246	0.56	66.51	0.58	72.64	73.22	0.0014	0.1733
15	−20.75	0.0108	2.4835	4.05	68.13	4.08	70.95	75.03	0.0095	0.1711
20	−8.13	0.0109	1.8977	6.73	69.37	6.77	69.63	76.40	0.0155	0.1697
30	11.11	0.0112	1.2964	10.86	71.25	10.93	67.53	78.45	0.0245	0.1679
40	25.93	0.0114	0.9874	14.08	72.69	14.16	65.84	80.00	0.0312	0.1668
50	38.15	0.0116	0.7982	16.75	73.86	16.86	64.39	81.25	0.0366	0.1660
60	48.64	0.0117	0.6701	19.07	74.86	19.20	63.10	82.30	0.0413	0.1654
70	57.90	0.0119	0.5772	21.13	75.73	21.29	61.92	83.21	0.0453	0.1649
80	66.21	0.0120	0.5068	23.00	76.50	23.18	60.82	84.00	0.0489	0.1645
90	73.79	0.0122	0.4514	24.72	77.20	24.92	59.79	84.71	0.0521	0.1642
100	80.76	0.0123	0.4067	26.31	77.82	26.54	58.81	85.35	0.0551	0.1639
120	93.29	0.0126	0.3389	29.21	78.93	29.49	56.97	86.46	0.0604	0.1634
140	104.35	0.0128	0.2896	31.82	79.89	32.15	55.24	87.39	0.0651	0.1630
160	114.30	0.0130	0.2522	34.21	80.71	34.59	53.59	88.18	0.0693	0.1626
180	123.38	0.0133	0.2228	36.42	81.44	36.86	52.00	88.86	0.0731	0.1623
200	131.74	0.0135	0.1989	38.50	82.08	39.00	50.44	89.44	0.0767	0.1620
220	139.51	0.0137	0.1792	40.48	82.08	41.03	48.90	89.94	0.0816	0.1616
240	146.77	0.0140	0.1625	42.35	83.14	42.97	47.39	90.36	0.0831	0.1613
260	153.60	0.0142	0.1483	44.16	83.58	44.84	45.88	90.72	0.0861	0.1609
280	160.06	0.0145	0.1359	45.90	83.97	46.65	44.36	91.01	0.0890	0.1605
300	166.18	0.0147	0.1251	47.59	84.30	48.41	42.83	91.24	0.0917	0.1601

TABLE A-13E
Superheated refrigerant-12

Temp. °F	v ft³/lbm	u Btu/lbm	h Btu/lbm	s Btu/(lbm·R)	v ft³/lbm	u Btu/lbm	h Btu/lbm	s Btu/(lbm·R)
	10 psia ($T_{sat} = -37.23°F$)				**15 psia ($T_{sat} = -20.75°F$)**			
Sat.	3.6246	66.512	73.219	0.1733	2.4835	68.134	75.028	0.1711
0	3.9809	70.879	78.246	0.1847	2.6201	70.629	77.902	0.1775
20	4.1691	73.299	81.104	0.1906	2.7494	73.080	80.712	0.1835
40	4.3556	75.768	83.828	0.1964	2.8770	75.575	83.561	0.1893
60	4.5408	78.286	86.689	0.2020	3.0031	78.115	86.451	0.1950
80	4.7248	80.853	89.596	0.2075	3.1281	80.700	89.383	0.2005
100	4.9079	83.466	92.548	0.2128	3.2521	83.330	92.357	0.2059
120	5.0903	86.126	95.546	0.2181	3.3754	86.004	95.373	0.2112
140	5.2720	88.830	98.586	0.2233	3.4981	88.719	98.429	0.2164
160	5.4533	91.578	101.669	0.2283	3.6202	91.476	101.525	0.2215
180	5.6341	94.367	104.793	0.2333	3.7419	94.274	104.661	0.2265
200	5.8145	97.197	107.957	0.2381	3.8632	97.112	107.835	0.2314
	20 psia ($T_{sat} = -8.13°F$)				**30 psia ($T_{sat} = 11.11°F$)**			
Sat.	1.8977	69.374	76.397	0.1697	1.2964	71.255	78.452	0.1679
20	2.0391	72.856	80.403	0.1783	1.3278	72.394	79.765	0.1707
40	2.1373	75.379	83.289	0.1842	1.3969	74.975	82.730	0.1767
60	2.2340	77.942	86.210	0.1899	1.4644	77.586	85.716	0.1826
80	2.3295	80.546	89.168	0.1955	1.5306	80.232	88.729	0.1883
100	2.4241	83.192	92.164	0.2010	1.5957	82.911	91.770	0.1938
120	2.5179	85.879	95.198	0.2063	1.6600	85.627	94.843	0.1992
140	2.6110	88.607	98.270	0.2115	1.7237	88.379	97.948	0.2045
160	2.7036	91.374	101.380	0.2166	1.7868	91.166	101.086	0.2096
180	2.7957	94.181	104.528	0.2216	1.8494	93.991	104.258	0.2146
200	2.8874	97.026	107.712	0.2265	1.9116	96.852	107.464	0.2196
220	2.9789	99.907	110.932	0.2313	1.9735	99.746	110.702	0.2244
	40 psia ($T_{sat} = 25.93°F$)				**50 psia ($T_{sat} = 38.15°F$)**			
Sat.	0.9874	72.691	80.000	0.1668	0.7982	73.863	81.249	0.1660
40	1.0258	74.555	82.148	0.1711	0.8025	74.115	81.540	0.1666
60	1.0789	77.220	85.206	0.1771	0.8471	76.838	84.676	0.1727
80	1.1306	79.908	88.277	0.1829	0.8903	79.574	87.811	0.1786
100	1.1812	82.624	91.367	0.1885	0.9322	82.328	90.953	0.1843
120	1.2309	85.369	94.480	0.1940	0.9731	85.106	94.110	0.1899
140	1.2798	88.147	97.620	0.1993	1.0133	87.910	97.286	0.1953
160	1.3282	90.957	100.788	0.2045	1.0529	90.743	100.485	0.2005
180	1.3761	93.800	103.985	0.2096	1.0920	93.604	103.708	0.2056
200	1.4236	96.674	107.212	0.2146	1.1307	96.496	106.958	0.2106
220	1.4707	99.583	110.469	0.2194	1.1690	99.419	110.235	0.2155
240	1.5176	102.524	113.757	0.2242	1.2070	102.371	113.539	0.2203

Temp. °F	v ft³/lbm	u Btu/lbm	h Btu/lbm	s Btu/(lbm · R)	v ft³/lbm	u Btu/lbm	h Btu/lbm	s Btu/(lbm
	\multicolumn{4}{c	}{**60 psia** ($T_{sat} = 48.64°F$)}	\multicolumn{4}{c}{**70 psia** ($T_{sat} = 57.90°F$)}			
Sat.	0.6701	74.859	82.299	0.1654	0.5772	75.729	83.206	0.1649
60	0.6921	76.442	84.126	0.1689	0.5809	76.027	83.552	0.1656
80	0.7296	79.229	87.330	0.1750	0.6146	78.871	86.832	0.1718
100	0.7659	82.024	90.528	0.1808	0.6469	81.712	90.091	0.1777
120	0.8011	84.836	93.731	0.1864	0.6780	84.560	93.343	0.1834
140	0.8335	87.668	96.945	0.1919	0.7084	87.421	96.597	0.1889
160	0.8693	90.524	100.776	0.1972	0.7380	90.302	99.862	0.1943
180	0.9025	93.406	103.427	0.2023	0.7671	93.205	103.141	0.1995
200	0.9353	96.315	106.700	0.2074	0.7957	96.132	106.439	0.2046
220	0.9678	99.252	109.997	0.2123	0.8240	99.083	109.756	0.2095
240	0.9998	102.217	113.319	0.2171	0.8519	102.061	113.096	0.2144
260	1.0318	105.210	116.666	0.2218	0.8796	105.065	116.459	0.2191
	\multicolumn{4}{c	}{**80 psia** ($T_{sat} = 66.21°F$)}	\multicolumn{4}{c}{**90 psia** ($T_{sat} = 73.79°F$)}			
Sat.	0.5068	76.500	84.003	0.1645	0.4514	77.194	84.713	0.1642
80	0.5280	78.500	86.316	0.1689	0.4602	78.115	85.779	0.1662
100	0.5573	81.389	89.640	0.1749	0.4875	81.056	89.175	0.1723
120	0.5856	84.276	92.945	0.1807	0.5135	83.984	92.536	0.1782
140	0.6129	87.169	96.242	0.1863	0.5385	86.911	95.879	0.1839
160	0.6394	90.076	99.542	0.1917	0.5627	89.845	99.216	0.1894
180	0.6654	93.000	102.851	0.1970	0.5863	92.793	102.557	0.1947
200	0.6910	95.945	106.174	0.2021	0.6094	95.755	105.905	0.1998
220	0.7161	98.912	109.513	0.2071	0.6321	98.739	109.267	0.2049
240	0.7409	101.904	112.872	0.2119	0.6545	101.743	112.644	0.2098
260	0.7654	104.919	116.251	0.2167	0.6766	104.771	116.040	0.2146
280	0.7898	107.960	119.652	0.2214	0.6985	107.823	119.456	0.2192
	\multicolumn{4}{c	}{**100 psia** ($T_{sat} = 80.76°F$)}	\multicolumn{4}{c}{**120 psia** ($T_{sat} = 93.29°F$)}			
Sat.	0.4067	77.824	85.351	0.1639	0.3389	78.933	86.459	0.1634
100	0.4314	80.711	88.694	0.1700	0.3466	79.978	87.675	0.1656
120	0.4556	83.685	92.116	0.1760	0.3684	83.056	91.237	0.1718
140	0.4788	86.647	95.507	0.1817	0.3890	86.098	94.736	0.1778
160	0.5012	89.610	98.884	0.1873	0.4087	89.123	98.199	0.1835
180	0.5229	92.580	102.257	0.1926	0.4277	92.144	101.642	0.1889
200	0.5441	95.564	105.633	0.1978	0.4461	95.170	105.076	0.1942
220	0.5649	98.564	109.018	0.2029	0.4640	98.205	108.509	0.1993
240	0.5854	101.582	112.415	0.2078	0.4816	101.253	111.948	0.2043
260	0.6055	104.622	115.828	0.2126	0.4989	104.317	115.396	0.2092
280	0.6255	107.684	119.258	0.2173	0.5159	107.401	118.857	0.2139
300	0.6452	110.768	122.707	0.2219	0.5327	110.504	122.333	0.2186

	v ft³/lbm	u Btu/lbm	h Btu/lbm	s Btu/(lbm·R)	v ft³/lbm	u Btu/lbm	h Btu/lbm	s Btu/(lbm·R)
	140 psia (T_{sat} = **104.35°F**)				**160 psia** (T_{sat} = **114.30°F**)			
		79.886	87.389	0.1630	0.2522	80.713	88.180	0.1626
		82.382	90.297	0.1681	0.2576	81.656	89.283	0.1645
		85.516	93.923	0.1742	0.2756	84.899	93.059	0.1709
	…3	88.615	97.483	0.1801	0.2922	88.080	96.732	0.1770
	…594	91.692	101.003	0.1857	0.3080	91.221	100.340	0.1827
	…3758	94.765	104.501	0.1910	0.3230	94.344	103.907	0.1882
	0.3918	97.837	107.987	0.1963	0.3375	97.457	107.450	0.1935
	0.4073	100.918	111.470	0.2013	0.3516	100.570	110.980	0.1986
	0.4226	104.008	114.956	0.2062	0.3653	103.690	114.506	0.2036
	0.4375	107.115	118.449	0.2110	0.3787	106.820	118.033	0.2084
	0.4523	110.235	121.953	0.2157	0.3919	109.964	121.567	0.2131
	0.4668	113.376	125.470	0.2202	0.4049	113.121	125.109	0.2177
	180 psia (T_{sat} = **123.38°F**)				**200 psia** (T_{sat} = **131.74°F**)			
Sat.	0.2228	81.436	88.857	0.1623	0.1989	82.077	89.439	0.1620
140	0.2371	84.238	92.136	0.1678	0.2058	83.521	91.137	0.1648
160	0.2530	87.513	95.940	0.1741	0.2212	86.913	95.100	0.1713
180	0.2678	90.727	99.647	0.1800	0.2354	90.211	98.921	0.1774
200	0.2818	93.904	103.291	0.1856	0.2486	93.451	102.652	0.1831
220	0.2952	97.063	106.896	0.1910	0.2612	96.659	106.325	0.1886
240	0.3081	100.215	110.478	0.1961	0.2732	99.850	109.962	0.1939
260	0.3207	103.364	114.046	0.2012	0.2849	103.032	113.576	0.1990
280	0.3329	106.521	117.610	0.2061	0.2962	106.214	117.178	0.2039
300	0.3449	109.686	121.174	0.2108	0.3073	109.402	120.775	0.2087
320	0.3567	112.863	124.744	0.2155	0.3182	112.598	124.373	0.2134
340	0.3683	116.053	128.321	0.2200	0.3288	115.805	127.974	0.2179
	300 psia (T_{sat} = **166.18°F**)				**400 psia** (T_{sat} = **192.93°F**)			
Sat.	0.1251	84.295	91.240	0.1601	0.0856	85.178	91.513	0.1576
180	0.1348	87.071	94.556	0.1654				
200	0.1470	90.816	98.975	0.1722	0.0910	86.982	93.718	0.1609
220	0.1577	94.379	103.136	0.1784	0.1032	91.410	99.046	0.1689
240	0.1676	97.835	107.140	0.1842	0.1130	95.371	103.735	0.1757
260	0.1769	101.225	111.043	0.1897	0.1216	99.102	108.105	0.1818
280	0.1856	104.574	114.879	0.1950	0.1295	102.701	112.286	0.1876
300	0.1940	107.899	118.670	0.2000	0.1368	106.217	116.343	0.1930
320	0.2021	111.208	122.430	0.2049	0.1437	109.680	120.318	0.1981
340	0.2100	114.512	126.171	0.2096	0.1503	113.108	124.235	0.2031
360	0.2177	117.814	129.900	0.2142	0.1567	116.514	128.112	0.2079

P-h diagram for refrigerant-12. (Freon 12 is the du Pont trademark for refriger
Copyright E. I. du Pont de Nemours & Company; used with permis

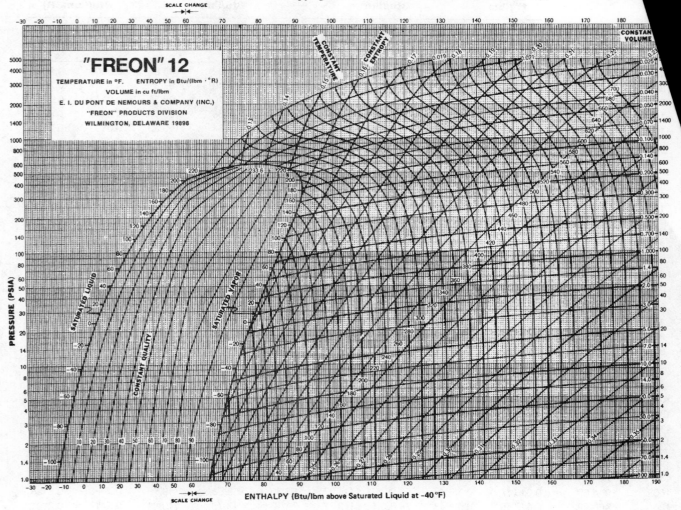

ENTHALPY (Btu/lbm above Saturated Liquid at -40°F)

949

perature table

		Specific volume ft³/lbm		Internal energy Btu/lbm		Enthalpy Btu/lbm			Entropy Btu/(lbm · R)	
		Sat. liquid v_f	Sat. vapor v_g	Sat. liquid u_f	Sat. vapor u_g	Sat. liquid h_f	Evap. h_{fg}	Sat. vapor h_g	Sat. liquid s_f	Sat. vapor s_g
		0.011 30	5.7173	−0.02	87.90	0.00	95.82	95.82	0.0000	0.2283
	0	0.011 43	4.3911	2.81	89.26	2.83	94.49	97.32	0.0067	0.2266
	949	0.011 56	3.4173	5.69	90.62	5.71	93.10	98.81	0.0133	0.2250
	4.718	0.011 63	3.0286	7.14	91.30	7.17	92.38	99.55	0.0166	0.2243
	16.674	0.011 70	2.6918	8.61	91.98	8.65	91.64	100.29	0.0199	0.2236
	18.831	0.011 78	2.3992	10.09	92.66	10.13	90.89	101.02	0.0231	0.2230
	21.203	0.011 85	2.1440	11.58	93.33	11.63	90.12	101.75	0.0264	0.2224
	23.805	0.011 93	1.9208	13.09	94.01	13.14	89.33	102.47	0.0296	0.2219
	26.651	0.012 00	1.7251	14.60	94.68	14.66	88.53	103.19	0.0329	0.2214
5	29.756	0.012 08	1.5529	16.13	95.35	16.20	87.71	103.90	0.0361	0.2209
20	33.137	0.012 16	1.4009	17.67	96.02	17.74	86.87	104.61	0.0393	0.2205
25	36.809	0.012 25	1.2666	19.22	96.69	19.30	86.02	105.32	0.0426	0.2200
30	40.788	0.012 33	1.1474	20.78	97.35	20.87	85.14	106.01	0.0458	0.2196
40	49.738	0.012 51	0.9470	23.94	98.67	24.05	83.34	107.39	0.0522	0.2189
50	60.125	0.012 70	0.7871	27.14	99.98	27.28	81.46	108.74	0.0585	0.2183
60	72.092	0.012 90	0.6584	30.39	101.27	30.56	79.49	110.05	0.0648	0.2178
70	85.788	0.013 11	0.5538	33.68	102.54	33.89	77.44	111.33	0.0711	0.2173
80	101.37	0.013 34	0.4682	37.02	103.78	37.27	75.29	112.56	0.0774	0.2169
85	109.92	0.013 46	0.4312	38.72	104.39	38.99	74.17	113.16	0.0805	0.2167
90	118.99	0.013 58	0.3975	40.42	105.00	40.72	73.03	113.75	0.0836	0.2165
95	128.62	0.013 71	0.3668	42.14	105.60	42.47	71.86	114.33	0.0867	0.2163
100	138.83	0.013 85	0.3388	43.87	106.18	44.23	70.66	114.89	0.0898	0.2161
105	149.63	0.013 99	0.3131	45.62	106.76	46.01	69.42	115.43	0.0930	0.2159
110	161.04	0.014 14	0.2896	47.39	107.33	47.81	68.15	115.96	0.0961	0.2157
115	173.10	0.014 29	0.2680	49.17	107.88	49.63	66.84	116.47	0.0992	0.2155
120	185.82	0.014 45	0.2481	50.97	108.42	51.47	65.48	116.95	0.1023	0.2153
140	243.86	0.015 20	0.1827	58.39	110.41	59.08	59.57	118.65	0.1150	0.2143
160	314.63	0.016 17	0.1341	66.26	111.97	67.20	52.58	119.78	0.1280	0.2128
180	400.22	0.017 58	0.0964	74.83	112.77	76.13	43.78	119.91	0.1417	0.2101
200	503.52	0.020 14	0.0647	84.90	111.66	86.77	30.92	117.69	0.1575	0.2044
210	563.51	0.023 29	0.0476	91.84	108.48	94.27	19.18	113.45	0.1684	0.1971

Source: Tables A-15E and A-16E are adopted from M. J. Moran and H. N. Shapiro, Fundamentals of Engineering Thermodynamics, 2d ed., Wiley, New York, 1992, pp. 754–758. Originally based on equations from D. P. Wilson and R. S. Basu, "Thermodynamic Properties of a New Stratospherically Safe Working Fluid—Refrigerant 134a," ASHRAE Trans., Vol. 94, Pt. 2, 1988, pp. 2095–2118.

Press. psia P	Temp. °F T_{sat}	Specific volume ft³/lbm		Internal energy Btu/lbm		Enthalpy Btu/lbm			Entropy Btu/(lbm·R)	
		Sat. liquid v_f	Sat. vapor v_g	Sat. liquid u_f	Sat. vapor u_g	Sat. liquid h_f	Evap. h_{fg}	Sat. vapor h_g	Sat. liquid s_f	Sat. vapor s_g
5	−53.48	0.011 13	8.3508	−3.74	86.07	−3.73	97.53	93.79	−0.0090	0.2311
10	−29.71	0.011 43	4.3581	2.89	89.30	2.91	94.45	97.37	0.0068	0.2265
15	−14.25	0.011 64	2.9747	7.36	91.40	7.40	92.27	99.66	0.0171	0.2242
20	−2.48	0.011 81	2.2661	10.84	93.00	10.89	90.50	101.39	0.0248	0.2227
30	15.38	0.012 09	1.5408	16.24	95.40	16.31	87.65	103.96	0.0364	0.2209
40	29.04	0.012 32	1.1692	20.48	97.23	20.57	85.31	105.88	0.0452	0.2197
50	40.27	0.012 52	0.9422	24.02	98.71	24.14	83.29	107.43	0.0523	0.2189
60	49.89	0.012 70	0.7887	27.10	99.96	27.24	81.48	108.72	0.0584	0.2183
70	58.35	0.012 86	0.6778	29.85	101.05	30.01	79.82	109.83	0.0638	0.2179
80	65.93	0.013 02	0.5938	32.33	102.02	32.53	78.28	110.81	0.0686	0.2175
90	72.83	0.013 17	0.5278	34.62	102.89	34.84	76.84	111.68	0.0729	0.2172
100	79.17	0.013 32	0.4747	36.75	103.68	36.99	75.47	112.46	0.0768	0.2169
120	90.54	0.013 60	0.3941	40.61	105.06	40.91	72.91	113.82	0.0839	0.2165
140	100.56	0.013 86	0.3358	44.07	106.25	44.43	70.52	114.95	0.0902	0.2161
160	109.56	0.014 12	0.2916	47.23	107.28	47.65	68.26	115.91	0.0958	0.2157
180	117.74	0.014 38	0.2569	50.16	108.18	50.64	66.10	116.74	0.1009	0.2154
200	125.28	0.014 63	0.2288	52.90	108.98	53.44	64.01	117.44	0.1057	0.2151
220	132.27	0.014 89	0.2056	55.48	109.68	56.09	61.96	118.05	0.1101	0.2147
240	138.79	0.015 15	0.1861	57.93	110.30	58.61	59.96	118.56	0.1142	0.2144
260	144.92	0.015 41	0.1695	60.28	110.84	61.02	57.97	118.99	0.1181	0.2140
280	150.70	0.015 68	0.1550	62.53	111.31	63.34	56.00	119.35	0.1219	0.2136
300	156.17	0.015 96	0.1424	64.71	111.72	65.59	54.03	119.62	0.1254	0.2132
350	168.72	0.016 71	0.1166	69.88	112.45	70.97	49.03	120.00	0.1338	0.2118
400	179.95	0.017 58	0.0965	74.81	112.77	76.11	43.80	119.91	0.1417	0.2102
450	190.12	0.018 63	0.0800	79.63	112.60	81.18	38.08	119.26	0.1493	0.2079
500	199.38	0.020 02	0.0657	84.54	111.76	86.39	31.44	117.83	0.1570	0.2047

TABLE A-16E
Superheated refrigerant-134a

T °F	v ft³/lbm	u Btu/lbm	h Btu/lbm	s Btu/(lbm·R)	v ft³/lbm	u Btu/lbm	h Btu/lbm	s Btu/(lbm·R)
	$P = 10$ psia ($T_{sat} = -29.71°F$)				$P = 15$ psia ($T_{sat} = -14.25°F$)			
Sat.	4.3581	89.30	97.37	0.2265	2.9747	91.40	99.66	0.2242
−20	4.4718	90.89	99.17	0.2307				
0	4.7026	94.24	102.94	0.2391	3.0893	93.84	102.42	0.2303
20	4.9297	97.67	106.79	0.2472	3.2468	97.33	106.34	0.2386
40	5.1539	101.19	110.72	0.2553	3.4012	100.89	110.33	0.2468
60	5.3758	104.80	114.74	0.2632	3.5533	104.54	114.40	0.2548
80	5.5959	108.50	118.85	0.2709	3.7034	108.28	118.56	0.2626
100	5.8145	112.29	123.05	0.2786	3.8520	112.10	122.79	0.2703
120	6.0318	116.18	127.34	0.2861	3.9993	116.01	127.11	0.2779
140	6.2482	120.16	131.72	0.2935	4.1456	120.00	131.51	0.2854
160	6.4638	124.23	136.19	0.3009	4.2911	124.09	136.00	0.2927
180	6.6786	128.38	140.74	0.3081	4.4359	128.26	140.57	0.3000
200	6.8929	132.63	145.39	0.3152	4.5801	132.52	145.23	0.3072
	$P = 20$ psia ($T_{sat} = -2.48°F$)				$P = 30$ psia ($T_{sat} = 15.38°F$)			
Sat.	2.2661	93.00	101.39	0.2227	1.5408	95.40	103.96	0.2209
0	2.2816	93.43	101.88	0.2238				
20	2.4046	96.98	105.88	0.2323	1.5611	96.26	104.92	0.2229
40	2.5244	100.59	109.94	0.2406	1.6465	99.98	109.12	0.2315
60	2.6416	104.28	114.06	0.2487	1.7293	103.75	113.35	0.2398
80	2.7569	108.05	118.25	0.2566	1.8098	107.59	117.63	0.2478
100	2.8705	111.90	122.52	0.2644	1.8887	111.49	121.98	0.2558
120	2.9829	115.83	126.87	0.2720	1.9662	115.47	126.39	0.2635
140	3.0942	119.85	131.30	0.2795	2.0426	119.53	130.87	0.2711
160	3.2047	123.95	135.81	0.2869	2.1181	123.66	135.42	0.2786
180	3.3144	128.13	140.40	0.2922	2.1929	127.88	140.05	0.2859
200	3.4236	132.40	145.07	0.3014	2.2671	132.17	144.76	0.2932
220	3.5323	136.76	149.83	0.3085	2.3407	136.55	149.54	0.3003
	$P = 40$ psia ($T_{sat} = 29.04°F$)				$P = 50$ psia ($T_{sat} = 40.27°F$)			
Sat.	1.1692	97.23	105.88	0.2197	0.9422	98.71	107.43	0.2189
40	1.2065	99.33	108.26	0.2245				
60	1.2723	103.20	112.62	0.2331	0.9974	102.62	111.85	0.2276
80	1.3357	107.11	117.00	0.2414	1.0508	106.62	116.34	0.2361
100	1.3973	111.08	121.42	0.2494	1.1022	110.65	120.85	0.2443
120	1.4575	115.11	125.90	0.2573	1.1520	114.74	125.39	0.2523
140	1.5165	119.21	130.43	0.2650	1.2007	118.88	129.99	0.2601
160	1.5746	123.38	135.03	0.2725	1.2484	123.08	134.64	0.2677
180	1.6319	127.62	139.70	0.2799	1.2953	127.36	139.34	0.2752
200	1.6887	131.94	144.44	0.2872	1.3415	131.71	144.12	0.2825
220	1.7449	136.34	149.25	0.2944	1.3873	136.12	148.96	0.2897
240	1.8006	140.81	154.14	0.3015	1.4326	140.61	153.87	0.2969
260	1.8561	145.36	159.10	0.3085	1.4775	145.18	158.85	0.3039
280	1.9112	149.98	164.13	0.3154	1.5221	149.82	163.90	0.3108

T °F	v ft³/lbm	u Btu/lbm	h Btu/lbm	s Btu/(lbm · R)	v ft³/lbm	u Btu/lbm	h Btu/lbm	s Btu/(lbm · R)
	P = 60 psia (T_{sat} = 49.89°F)				**P = 70 psia (T_{sat} = 58.35°F)**			
Sat.	0.7887	99.96	108.72	0.2183	0.6778	101.05	109.83	0.2179
60	0.8135	102.03	111.06	0.2229	0.6814	101.40	110.23	0.2186
80	0.8604	106.11	115.66	0.2316	0.7239	105.58	114.96	0.2276
100	0.9051	110.21	120.26	0.2399	0.7640	109.76	119.66	0.2361
120	0.9482	114.35	124.88	0.2480	0.8023	113.96	124.36	0.2444
140	0.9900	118.54	129.53	0.2559	0.8393	118.20	129.07	0.2524
160	1.0308	122.79	134.23	0.2636	0.8752	122.49	133.82	0.2601
180	1.0707	127.10	138.98	0.2712	0.9103	126.83	138.62	0.2678
200	1.1100	131.47	143.79	0.2786	0.9446	131.23	143.46	0.2752
220	1.1488	135.91	148.66	0.2859	0.9784	135.69	148.36	0.2825
240	1.1871	140.42	153.60	0.2930	1.0118	140.22	153.33	0.2897
260	1.2251	145.00	158.60	0.3001	1.0448	144.82	158.35	0.2968
280	1.2627	149.65	163.67	0.3070	1.0774	149.48	163.44	0.3038
300	1.3001	154.38	168.81	0.3139	1.1098	154.22	168.60	0.3107
	P = 80 psia (T_{sat} = 65.93°F)				**P = 90 psia (T_{sat} = 72.83°F)**			
Sat.	0.5938	102.02	110.81	0.2175	0.5278	102.89	111.68	0.2172
80	0.6211	105.03	114.23	0.2239	0.5408	104.46	113.47	0.2205
100	0.6579	109.30	119.04	0.2327	0.5751	108.82	118.39	0.2295
120	0.6927	113.56	123.82	0.2411	0.6073	113.15	123.27	0.2380
140	0.7261	117.85	128.60	0.2492	0.6380	117.50	128.12	0.2463
160	0.7584	122.18	133.41	0.2570	0.6675	121.87	132.98	0.2542
180	0.7898	126.55	138.25	0.2647	0.6961	126.28	137.87	0.2620
200	0.8205	130.98	143.13	0.2722	0.7239	130.73	142.79	0.2696
220	0.8506	135.47	148.06	0.2796	0.7512	135.25	147.76	0.2770
240	0.8803	140.02	153.05	0.2868	0.7779	139.82	152.77	0.2843
260	0.9095	144.63	158.10	0.2940	0.8043	144.45	157.84	0.2914
280	0.9384	149.32	163.21	0.3010	0.8303	149.15	162.97	0.2984
300	0.9671	154.06	168.38	0.3079	0.8561	153.91	168.16	0.3054
320	0.9955	158.88	173.62	0.3147	0.8816	158.73	173.42	0.3122
	P = 100 psia (T_{sat} = 79.17°F)				**P = 120 psia (T_{sat} = 90.54°F)**			
Sat.	0.4747	103.68	112.46	0.2169	0.3941	105.06	113.82	0.2165
80	0.4761	103.87	112.68	0.2173				
100	0.5086	108.32	117.73	0.2265	0.4080	107.26	116.32	0.2210
120	0.5388	112.73	122.70	0.2352	0.4355	111.84	121.52	0.2301
140	0.5674	117.13	127.63	0.2436	0.4610	116.37	126.61	0.2387
160	0.5947	121.55	132.55	0.2517	0.4852	120.89	131.66	0.2470
180	0.6210	125.99	137.49	0.2595	0.5082	125.42	136.70	0.2550
200	0.6466	130.48	142.45	0.2671	0.5305	129.97	141.75	0.2628
220	0.6716	135.02	147.45	0.2746	0.5520	134.56	146.82	0.2704
240	0.6960	139.61	152.49	0.2819	0.5731	139.20	151.92	0.2778
260	0.7201	144.26	157.59	0.2891	0.5937	143.89	157.07	0.2850
280	0.7438	148.98	162.74	0.2962	0.6140	148.63	162.26	0.2921
300	0.7672	153.75	167.95	0.3031	0.6339	153.43	167.51	0.2991
320	0.7904	158.59	173.21	0.3099	0.6537	158.29	172.81	0.3060

TABLE A-16E
(Continued)

T °F	v ft³/lbm	u Btu/lbm	h Btu/lbm	s Btu/(lbm · R)	v ft³/lbm	u Btu/lbm	h Btu/lbm	s Btu/(lbm · R)
	\multicolumn{4}{c}{$P = $ **140 psia** ($T_{sat} = $ **100.56°F**)}	\multicolumn{4}{c}{$P = $ **160 psia** ($T_{sat} = $ **109.55°F**)}
Sat.	0.3358	106.25	114.95	0.2161	0.2916	107.28	115.91	0.2157
120	0.3610	110.90	120.25	0.2254	0.3044	109.88	118.89	0.2209
140	0.3846	115.58	125.24	0.2344	0.3269	114.73	124.41	0.2303
160	0.4066	120.21	130.74	0.2429	0.3474	119.49	129.78	0.2391
180	0.4274	124.82	135.89	0.2511	0.3666	124.20	135.06	0.2475
200	0.4474	129.44	141.03	0.2590	0.3849	128.90	140.29	0.2555
220	0.4666	134.09	146.18	0.2667	0.4023	133.61	145.52	0.2633
240	0.4852	138.77	151.34	0.2742	0.4192	138.34	150.75	0.2709
260	0.5034	143.50	156.54	0.2815	0.4356	143.11	156.00	0.2783
280	0.5212	148.28	161.78	0.2887	0.4516	147.92	161.29	0.2856
300	0.5387	153.11	167.06	0.2957	0.4672	152.78	166.61	0.2927
320	0.5559	157.99	172.39	0.3026	0.4826	157.69	171.98	0.2996
340	0.5730	162.93	177.78	0.3094	0.4978	162.65	177.39	0.3065
360	0.5898	167.93	183.21	0.3162	0.5128	167.67	182.85	0.3132
	\multicolumn{4}{c}{$P = $ **180 psia** ($T_{sat} = $ **117.74°F**)}	\multicolumn{4}{c}{$P = $ **200 psia** ($T_{sat} = $ **125.28°F**)}
Sat.	0.2569	108.18	116.74	0.2154	0.2288	108.98	117.44	0.2151
120	0.2595	108.77	117.41	0.2166				
140	0.2814	113.83	123.21	0.2264	0.2446	112.87	121.92	0.2226
160	0.3011	118.74	128.77	0.2355	0.2636	117.94	127.70	0.2321
180	0.3191	123.56	134.19	0.2441	0.2809	122.88	133.28	0.2410
200	0.3361	128.34	139.53	0.2524	0.2970	127.76	138.75	0.2494
220	0.3523	133.11	144.84	0.2603	0.3121	132.60	144.15	0.2575
240	0.3678	137.90	150.15	0.2680	0.3266	137.44	149.53	0.2653
260	0.3828	142.71	155.46	0.2755	0.3405	142.30	154.90	0.2728
280	0.3974	147.55	160.79	0.2828	0.3540	147.18	160.28	0.2802
300	0.4116	152.44	166.15	0.2899	0.3671	152.10	165.69	0.2874
320	0.4256	157.38	171.55	0.2969	0.3799	157.07	171.13	0.2945
340	0.4393	162.36	177.00	0.3038	0.3926	162.07	176.60	0.3014
360	0.4529	167.40	182.49	0.3106	0.4050	167.13	182.12	0.3082
	\multicolumn{4}{c}{$P = $ **300 psia** ($T_{sat} = $ **156.17°F**)}	\multicolumn{4}{c}{$P = $ **400 psia** ($T_{sat} = $ **179.95°F**)}
Sat.	0.1424	111.72	119.62	0.2132	0.0965	112.77	119.91	0.2102
160	0.1462	112.95	121.07	0.2155				
180	0.1633	118.93	128.00	0.2265	0.0965	112.79	119.93	0.2102
200	0.1777	124.47	134.34	0.2363	0.1143	120.14	128.60	0.2235
220	0.1905	129.79	140.36	0.2453	0.1275	126.35	135.79	0.2343
240	0.2021	134.99	146.21	0.2537	0.1386	132.12	142.38	0.2438
260	0.2130	140.12	151.95	0.2618	0.1484	137.65	148.64	0.2527
280	0.2234	145.23	157.63	0.2696	0.1575	143.06	154.72	0.2610
300	0.2333	150.33	163.28	0.2772	0.1660	148.39	160.67	0.2689
320	0.2428	155.44	168.92	0.2845	0.1740	153.69	166.57	0.2766
340	0.2521	160.57	174.56	0.2916	0.1816	158.97	172.42	0.2840
360	0.2611	165.74	180.23	0.2986	0.1890	164.26	178.26	0.2912
380	0.2699	170.94	185.92	0.3055	0.1962	169.57	184.09	0.2983
400	0.2786	176.18	191.64	0.3122	0.2032	174.90	189.94	0.3051

T R	h Btu/lbm	P_r	u Btu/lbm	v_r	s° Btu/(lbm·R)	T R	h Btu/lbm	P_r	u Btu/lbm	v_r	s° Btu/(lbm·R)
360	85.97	0.3363	61.29	396.6	0.503 69	1600	395.74	71.13	286.06	8.263	0.871 30
380	90.75	0.4061	64.70	346.6	0.516 63	1650	409.13	80.89	296.03	7.556	0.879 54
400	95.53	0.4858	68.11	305.0	0.528 90	1700	422.59	90.95	306.06	6.924	0.887 58
420	100.32	0.5760	71.52	270.1	0.540 58	1750	436.12	101.98	316.16	6.357	0.895 42
440	105.11	0.6776	74.93	240.6	0.551 72						
						1800	449.71	114.0	326.32	5.847	0.903 08
460	109.90	0.7913	78.36	215.33	0.562 35	1850	463.37	127.2	336.55	5.388	0.910 56
480	114.69	0.9182	81.77	193.65	0.572 55	1900	477.09	141.5	346.85	4.974	0.917 88
500	119.48	1.0590	85.20	174.90	0.582 33	1950	490.88	157.1	357.20	4.598	0.925 04
520	124.27	1.2147	88.62	158.58	0.591 73	2000	504.71	174.0	367.61	4.258	0.932 05
537	128.10	1.3593	91.53	146.34	0.599 45						
540	129.06	1.3860	92.04	144.32	0.600 78	2050	518.71	192.3	378.08	3.949	0.938 91
						2100	532.55	212.1	388.60	3.667	0.945 64
560	133.86	1.5742	95.47	131.78	0.609 50	2150	546.54	223.5	399.17	3.410	0.952 22
580	138.66	1.7800	98.90	120.70	0.617 93	2200	560.59	256.6	409.78	3.176	0.959 19
600	143.47	2.005	102.34	110.88	0.626 07	2250	574.69	281.4	420.46	2.961	0.965 01
620	148.28	2.249	105.78	102.12	0.633 95						
640	153.09	2.514	109.21	94.30	0.641 59	2300	588.82	308.1	431.16	2.765	0.971 23
						2350	603.00	336.8	441.91	2.585	0.977 32
660	157.92	2.801	112.67	87.27	0.649 02	2400	617.22	367.6	452.70	2.419	0.983 31
680	162.73	3.111	116.12	80.96	0.656 21	2450	631.48	400.5	463.54	2.266	0.989 19
700	167.56	3.446	119.58	75.25	0.663 21	2500	645.78	435.7	474.40	2.125	0.994 97
720	172.39	3.806	123.04	70.07	0.670 02						
740	177.23	4.193	126.51	65.38	0.676 65	2550	660.12	473.3	485.31	1.996	1.000 64
						2600	674.49	513.5	496.26	1.876	1.006 23
760	182.08	4.607	129.99	61.10	0.683 12	2650	688.90	556.3	507.25	1.765	1.011 72
780	186.94	5.051	133.47	57.20	0.689 42	2700	703.35	601.9	518.26	1.662	1.017 12
800	191.81	5.526	136.97	53.63	0.695 58	2750	717.83	650.4	529.31	1.566	1.022 44
820	196.69	6.033	140.47	50.35	0.701 60						
840	201.56	6.573	143.98	47.34	0.707 47	2800	732.33	702.0	540.40	1.478	1.027 67
						2850	746.88	756.7	551.52	1.395	1.032 82
860	206.46	7.149	147.50	44.57	0.713 23	2900	761.45	814.8	562.66	1.318	1.037 88
880	211.35	7.761	151.02	42.01	0.718 86	2950	776.05	876.4	573.84	1.247	1.042 88
900	216.26	8.411	154.57	39.64	0.724 38	3000	790.68	941.4	585.04	1.180	1.047 79
920	221.18	9.102	158.12	37.44	0.729 79						
940	226.11	9.834	161.68	35.41	0.735 09	3050	805.34	1011	596.28	1.118	1.052 64
						3100	820.03	1083	607.53	1.060	1.057 41
960	231.06	10.61	165.26	33.52	0.740 30	3150	834.75	1161	618.82	1.006	1.062 12
980	236.02	11.43	168.83	31.76	0.745 40	3200	849.48	1242	630.12	0.955	1.066 76
1000	240.98	12.30	172.43	30.12	0.750 42	3250	864.24	1328	641.46	0.907	1.071 34
1040	250.95	14.18	179.66	27.17	0.760 19						
1080	260.97	16.28	186.93	24.58	0.769 64	3300	879.02	1418	652.81	0.8621	1.075 85
						3350	893.83	1513	664.20	0.8202	1.080 31
1120	271.03	18.60	194.25	22.30	0.778 80	3400	908.66	1613	675.60	0.7807	1.084 70
1160	281.14	21.18	201.63	20.29	0.787 67	3450	923.52	1719	687.04	0.7436	1.089 04
1200	291.30	24.01	209.05	18.51	0.796 28	3500	938.40	1829	698.48	0.7087	1.093 32
1240	301.52	27.13	216.53	16.93	0.804 66						
1280	311.79	30.55	224.05	15.52	0.812 80	3550	953.30	1946	709.95	0.6759	1.097 55
						3600	968.21	2068	721.44	0.6449	1.101 72
1320	322.11	34.31	231.63	14.25	0.820 75	3650	983.15	2196	732.95	0.6157	1.105 84
1360	332.48	38.41	239.25	13.12	0.828 48	3700	998.11	2330	744.48	0.5882	1.109 91
1400	342.90	42.88	246.93	12.10	0.836 04	3750	1013.1	2471	756.04	0.5621	1.113 93
1440	353.37	47.75	254.66	11.17	0.843 41						
1480	363.89	53.04	262.44	10.34	0.850 62	3800	1028.1	2618	767.60	0.5376	1.117 91
						3850	1043.1	2773	779.19	0.5143	1.121 83
1520	374.47	58.78	270.26	9.578	0.857 67	3900	1058.1	2934	790.80	0.4923	1.125 71
1560	385.08	65.00	278.13	8.890	0.864 56	3950	1073.2	3103	802.43	0.4715	1.129 55

TABLE A-17E
(Continued)

T R	h Btu/lbm	P_r	u Btu/lbm	v_r	s° Btu/(lbm · R)	T R	h Btu/lbm	P_r	u Btu/lbm	v_r	s° Btu/(lbm · R)
4000	1088.3	3280	814.06	0.4518	1.133 34	4600	1270.4	6089	955.04	0.2799	1.175 75
						4700	1300.9	6701	978.73	0.2598	1.182 32
4050	1103.4	3464	825.72	0.4331	1.137 09	4800	1331.5	7362	1002.5	0.2415	1.188 76
4100	1118.5	3656	837.40	0.4154	1.140 79						
4150	1133.6	3858	849.09	0.3985	1.144 46	4900	1362.2	8073	1026.3	0.2248	1.195 08
4200	1148.7	4067	860.81	0.3826	1.148 09	5000	1392.9	8837	1050.1	0.2096	1.201 29
4300	1179.0	4513	884.28	0.3529	1.155 22	5100	1423.6	9658	1074.0	0.1956	1.207 38
						5200	1454.4	10,539	1098.0	0.1828	1.213 36
4400	1209.4	4997	907.81	0.3262	1.162 21	5300	1485.3	11,481	1122.0	0.1710	1.219 23
4500	1239.9	5521	931.39	0.3019	1.169 05						

Source: Kenneth Wark, *Thermodynamics,* 4th ed., McGraw-Hill, New York, 1983, pp. 832-833, table A-5. Originally published in J. H. Keenan and J. Kaye, *Gas Tables,* Wiley, New York, 1948.

T R	$\bar{h}$ Btu/lbmol	$\bar{u}$ Btu/lbmol	$\bar{s}°$ Btu/(lbmol · R)	T R	$\bar{h}$ Btu/lbmol	$\bar{u}$ Btu/lbmol	$\bar{s}°$ Btu/(lbmol · R)
300	2,082.0	1,486.2	41.695	1080	7,551.0	5,406.2	50.651
320	2,221.0	1,585.5	42.143	1100	7,695.0	5,510.5	50.783
340	2,360.0	1,684.4	42.564	1120	7,839.3	5,615.2	50.912
360	2,498.9	1,784.0	42.962	1140	7,984.0	5,720.1	51.040
380	2,638.0	1,883.4	43.337	1160	8,129.0	5,825.4	51.167
400	2,777.0	1,982.6	43.694	1180	8,274.4	5,931.0	51.291
420	2,916.1	2,082.0	44.034	1200	8,420.0	6,037.0	51.143
440	3,055.1	2,181.3	44.357	1220	8,566.1	6,143.4	51.534
460	3,194.1	2,280.6	44.665	1240	8,712.6	6,250.1	51.653
480	3,333.1	2,379.9	44.962	1260	8,859.3	6,357.2	51.771
500	3.472.2	2,479.3	45.246	1280	9,006.4	6,464.5	51.887
520	3.611.3	2,578.6	45.519	1300	9,153.9	6,572.3	51.001
537	3,729.5	2,663.1	45.743	1320	9,301.8	6,680.4	52.114
540	3,750.3	2,678.0	45.781	1340	9,450.0	6,788.9	52.225
560	3,889.5	2,777.4	46.034	1360	9,598.6	6,897.8	52.335
580	4,028.7	2,876.9	46.278	1380	9,747.5	7,007.0	52.444
600	4,167.9	2,976.4	46.514	1400	9,896.9	7,116.7	52.551
620	4,307.1	3,075.9	46.742	1420	10,046.6	7,226.7	52.658
640	4,446.4	3,175.5	46.964	1440	10,196.6	7,337.0	52.763
660	4,585.8	3,275.2	47.178	1460	10,347.0	7,447.6	52.867
680	4,725.3	3.374.9	47.386	1480	10,497.8	7,558.7	52.969
700	4,864.9	3,474.8	47.588	1500	10,648.0	7,670.1	53.071
720	5,004.5	3,574.7	47.785	1520	10,800.4	7,781.9	53.171
740	5,144.3	3,674.7	47.977	1540	10,952.2	7,893.9	53.271
760	5,284.1	3,774.9	48.164	1560	11,104.3	8,006.4	53.369
780	5,424.2	3,875.2	48.345	1580	11,256.9	8,119.2	53.465
800	5.564.4	3,975.7	48.522	1600	11,409.7	8,232.3	53.561
820	5,704.7	4,076.3	48.696	1620	11,562.8	8,345.7	53.656
840	5,845.3	4,177.1	48.865	1640	11,716.4	8,459.6	53.751
860	5,985.9	4,278.1	49.031	1660	11,870.2	8,573.6	53.844
880	6,126.9	4,379.4	49.193	1680	12,024.3	8,688.1	53.936
900	6,268.1	4,480.8	49.352	1700	12,178.9	8,802.9	54.028
920	6,409.6	4,582.6	49.507	1720	12,333.7	8,918.0	54.118
940	6,551.2	4,684.5	49.659	1740	12,488.8	9,033.4	54.208
960	6,693.1	4,786.7	49.808	1760	12,644.3	9,149.2	54.297
980	6,835.4	4,889.3	49.955	1780	12,800.2	9,265.3	54.385
1000	6,977.9	4,992.0	50.099	1800	12,956.3	9,381.7	54.472
1020	7,120.7	5,095.1	50.241	1820	13,112.7	9,498.4	54.559
1040	7,263.8	5,198.5	50.380	1840	13,269.5	9,615.5	54.645
1060	7,407.2	5,302.2	50.516	1860	13,426.5	9,732.8	54.729

TABLE A-18E
(Continued)

T R	$\bar{h}$ Btu/lbmol	$\bar{u}$ Btu/lbmol	$\bar{s}°$ Btu/(lbmol · R)	T R	$\bar{h}$ Btu/lbmol	$\bar{u}$ Btu/lbmol	$\bar{s}°$ Btu/(lbmol · R)
1900	13,742	9,968	54.896	3500	27,016	20,065	59.944
1940	14,058	10,205	55.061	3540	27,359	20,329	60.041
1980	14,375	10,443	55.223	3580	27,703	20,593	60.138
2020	14,694	10,682	55.383	3620	28,046	20,858	60.234
2060	15,013	10,923	55.540	3660	28,391	21,122	60.328
2100	15,334	11,164	55.694	3700	28,735	21,387	60.422
2140	15,656	11,406	55.846	3740	29,080	21,653	60.515
2180	15,978	11,649	55.995	3780	29,425	21,919	60.607
2220	16,302	11,893	56.141	3820	29,771	22,185	60.698
2260	16,626	12,138	56.286	3860	30,117	22,451	60.788
2300	16,951	12,384	56.429	3900	30,463	22,718	60.877
2340	17,277	12,630	56.570	3940	30,809	22,985	60.966
2380	17,604	12,878	56.708	3980	31,156	23,252	61.053
2420	17,392	13,126	56.845	4020	31,503	23,520	61.139
2460	18,260	13,375	56.980	4060	31,850	23,788	61.225
2500	18,590	13,625	57.112	4100	32,198	24,056	61.310
2540	18,919	13,875	57.243	4140	32,546	24,324	61.395
2580	19,250	14,127	57.372	4180	32,894	24,593	61.479
2620	19,582	14,379	57.499	4220	33,242	24,862	61.562
2660	19,914	14,631	57.625	4260	33,591	25,131	61.644
2700	20,246	14,885	57.750	4300	33,940	25,401	61.726
2740	20,580	15,139	57.872	4340	34,289	25,670	61.806
2780	20,914	15,393	57.993	4380	34,638	25,940	61.887
2820	21,248	15,648	58.113	4420	34,988	26,210	61.966
2860	21,584	15,905	58.231	4460	35,338	26,481	62.045
2900	21,920	16,161	58.348	4500	35,688	26,751	62.123
2940	22,256	16,417	58.463	4540	36,038	27,022	62.201
2980	22,593	16,675	58.576	4580	36,389	27,293	62.278
3020	22,930	16,933	58.688	4620	36,739	27,565	62.354
3060	23,268	17,192	58.800	4660	37,090	27,836	62.429
3100	23,607	17,451	58.910	4700	37,441	28,108	62.504
3140	23,946	17,710	59.019	4740	37,792	28,379	62.578
3180	24,285	17,970	59.126	4780	38,144	28,651	62.652
3220	24,625	18,231	59.232	4820	38,495	28,924	62.725
3260	24,965	18,491	59.338	4860	38,847	29,196	62.798
3300	25,306	18,753	59.442	4900	39,199	29,468	62.870
3340	25,647	19,014	59.544	5000	40,080	30,151	63.049
3380	25,989	19,277	59.646	5100	40,962	30,834	63.223
3420	26,331	19,539	59.747	5200	41,844	31,518	63.395
3460	26,673	19,802	59.846	5300	42,728	32,203	63.563

Source: Tables A-18E through A-23E are adapted from Kenneth Wark, *Thermodynamics,* 4th ed., McGraw-Hill, New York, 1983, pp. 834–844. Originally published in J. H. Keenan and J. Kaye, *Gas Tables,* Wiley, New York, 1945.

T R	$\bar{h}$ Btu/lbmol	$\bar{u}$ Btu/lbmol	$\bar{s}°$ Btu/(lbmol · R)	T R	$\bar{h}$ Btu/lbmol	$\bar{u}$ Btu/lbmol	$\bar{s}°$ Btu/(lbmol · R)
300	2,073.5	1,477.8	44.927	1080	7,696.8	5,552.1	54.064
320	2,212.6	1,577.1	45.375	1100	7,850.4	5,665.9	54.204
340	2,351.7	1,676.5	45.797	1120	8,004.5	5,780.3	54.343
360	2,490.8	1,775.9	46.195	1140	8,159.1	5,895.2	54.480
380	2,630.0	1,875.3	46.571	1160	8,314.2	6,010.6	54.614
400	2,769.1	1,974.8	46.927	1180	8,469.8	6,126.5	54.748
420	2,908.3	2,074.3	47.267	1200	8,625.8	6,242.8	54.879
440	3,047.5	2,173.8	47.591	1220	8,782.4	6,359.6	55.008
460	3,186.9	2,273.4	47.900	1240	8,939.4	6,476.9	55.136
480	3,326.5	2,373.3	48.198	1260	9,096.7	6,594.5	55.262
500	3,466.2	2,473.2	48.483	1280	9,254.6	6,712.7	55.386
520	3,606.1	2,573.4	48.757	1300	9,412.9	6,831.3	55.508
537	3,725.1	2,658.7	48.982	1320	9,571.9	6,950.2	55.630
540	3,746.2	2,673.8	49.021	1340	9,730.7	7,069.6	55.750
560	3,886.6	2,774.5	49.276	1360	9,890.2	7,189.4	55.867
580	4,027.3	2,875.5	49.522	1380	10,050.1	7,309.6	55.984
600	4,168.3	2,976.8	49.762	1400	10,210.4	7,430.1	56.099
620	4,309.7	3,078.4	49.993	1420	10,371.0	7,551.1	56.213
640	4,451.4	3,180.4	50.218	1440	10,532.0	7,672.4	56.326
660	4,593.5	3,282.9	50.437	1460	10,693.3	7,793.9	56.437
680	4,736.2	3,385.8	50.650	1480	10,855.1	7,916.0	56.547
700	4,879.3	3,489.2	50.858	1500	11,017.1	8,038.3	56.656
720	5,022.9	3,593.1	51.059	1520	11,179.6	8,161.1	56.763
740	5,167.0	3,697.4	51.257	1540	11,342.4	8,284.2	56.869
760	5,311.4	3,802.4	51.450	1560	11,505.4	8,407.4	56.975
780	5,456.4	3,907.5	51.638	1580	11,668.8	8,531.1	57.079
800	5,602.0	4,013.3	51.821	1600	11,832.5	8,655.1	57.182
820	5,748.1	4,119.7	52.002	1620	11,996.6	8,779.5	57.284
840	5,894.8	4,226.6	52.179	1640	12,160.9	8,904.1	57.385
860	6,041.9	4,334.1	52.352	1660	12,325.5	9,029.0	57.484
880	6,189.6	4,442.0	52.522	1680	12,490.4	9,154.1	57.582
900	6,337.9	4,550.6	52.688	1700	12,655.6	9,279.6	57.680
920	6,486.7	4,659.7	52.852	1720	12,821.1	9,405.4	57.777
940	6,636.1	4,769.4	53.012	1740	12,986.9	9,531.5	57.873
960	6,786.0	4,879.5	53.170	1760	13,153.0	9,657.9	57.968
980	6,936.4	4,990.3	53.326	1780	13,319.2	9,784.4	58.062
1000	7,087.5	5,101.6	53.477	1800	13,485.8	9,911.2	58.155
1020	7,238.9	5,213.3	53.628	1820	13,652.5	10,038.2	58.247
1040	7,391.0	5,325.7	53.775	1840	13,819.6	10,165.6	58.339
1060	7,543.6	5,438.6	53.921	1860	13,986.8	10,293.1	58.428

TABLE A-19E
(*Continued*)

T R	$\bar{h}$ Btu/lbmol	$\bar{u}$ Btu/lbmol	$\bar{s}°$ Btu/(lbmol · R)	T R	$\bar{h}$ Btu/lbmol	$\bar{u}$ Btu/lbmol	$\bar{s}°$ Btu/(lbmol · R)
1900	14,322	10,549	58.607	3500	28,273	21,323	63.914
1940	14,658	10,806	58.782	3540	28,633	21,603	64.016
1980	14,995	11,063	58.954	3580	28,994	21,884	64.114
2020	15,333	11,321	59.123	3620	29,354	22,165	64.217
2060	15,672	11,581	59.289	3660	29,716	22,447	64.316
2100	16,011	11,841	59.451	3700	30,078	22,730	64.415
2140	16,351	12,101	59.612	3740	30,440	23,013	64.512
2180	16,692	12,363	59.770	3780	30,803	23,296	64.609
2220	17,036	12,625	59.926	3820	31,166	23,580	64.704
2260	17,376	12,888	60.077	3860	31,529	23,864	64.800
2300	17,719	13,151	60.228	3900	31,894	24,149	64.893
2340	18,062	13,416	60.376	3940	32,258	24,434	64.986
2380	18,407	13,680	60.522	3980	32,623	24,720	65.078
2420	18,572	13,946	60.666	4020	32,989	25,006	65.169
2460	19,097	14,212	60.808	4060	33,355	25,292	65.260
2500	19,443	14,479	60.946	4100	33,722	25,580	65.350
2540	19,790	14,746	61.084	4140	34,089	25,867	64.439
2580	20,138	15,014	61.220	4180	34,456	26,155	65.527
2620	20,485	15,282	61.354	4220	34,824	26,144	65.615
2660	20,834	15,551	61.486	4260	35,192	26,733	65.702
2700	21,183	15,821	61.616	4300	35,561	27,022	65.788
2740	21,533	16,091	61.744	4340	35,930	27,312	65.873
2780	21,883	16,362	61.871	4380	36,300	27,602	65.958
2820	22,232	16,633	61.996	4420	36,670	27,823	66.042
2860	22,584	16,905	62.120	4460	37,041	28,184	66.125
2900	22,936	17,177	62.242	4500	37,412	28,475	66.208
2940	23,288	17,450	62.363	4540	37,783	28,768	66.290
2980	23,641	17,723	62.483	4580	38,155	29,060	66.372
3020	23,994	17,997	62.599	4620	38,528	29,353	66.453
3060	24,348	18,271	62.716	4660	38,900	29,646	66.533
3100	24,703	18,546	62.831	4700	39,274	29,940	66.613
3140	25,057	18,822	62.945	4740	39,647	30,234	66.691
3180	25,413	19,098	63.057	4780	40,021	30,529	66.770
3220	25,769	19,374	63.169	4820	40,396	30,824	66.848
3260	26,175	19,651	63.279	4860	40,771	31,120	66.925
3300	26,412	19,928	63.386	4900	41,146	31,415	67.003
3340	26,839	20,206	63.494	5000	42,086	32,157	67.193
3380	27,197	20,485	63.601	5100	43,021	32,901	67.380
3420	27,555	20,763	63.706	5200	43,974	33,648	67.562
3460	27,914	21,043	63.811	5300	44,922	34,397	67.743

T R	$\bar{h}$ Btu/lbmol	$\bar{u}$ Btu/lbmol	$\bar{s}°$ Btu/(lbmol · R)	T R	$\bar{h}$ Btu/lbmol	$\bar{u}$ Btu/lbmol	$\bar{s}°$ Btu/(lbmol · R)
300	2,108.2	1,512.4	46.353	1080	9,575.8	7,431.1	58.072
320	2,256.6	1,621.1	46.832	1100	9,802.6	7,618.1	58.281
340	2,407.3	1,732.1	47.289	1120	10,030.6	7,806.4	58.485
360	2,560.5	1,845.6	47.728	1140	10,260.1	7,996.2	58.689
380	2,716.4	1,961.8	48.148	1160	10,490.6	8,187.0	58.889
400	2,874.7	2,080.4	48.555	1180	10,722.3	8,379.0	59.088
420	3,035.7	2,201.7	48.947	1200	10,955.3	8,572.3	59.283
440	3,199.4	2,325.6	49.329	1220	11,189.4	8,766.6	59.477
460	3,365.7	2,452.2	49.698	1240	11,424.6	8,962.1	59.668
480	3,534.7	2,581.5	50.058	1260	11,661.0	9,158.8	59.858
500	3,706.2	2,713.3	50.408	1280	11,898.4	9,356.5	60.044
520	3,880.3	2,847.7	50.750	1300	12,136.9	9,555.3	60.229
537	4,027.5	2,963.8	51.032	1320	12,376.4	9,755.0	60.412
540	4,056.8	2,984.4	51.082	1340	12,617.0	9,955.9	60.593
560	4,235.8	3,123.7	51.408	1360	12,858.5	10,157.7	60.772
580	4,417.2	3,265.4	51.726	1380	13,101.0	10,360.5	60.949
600	4,600.9	3,409.4	52.038	1400	13,344.7	10,564.5	61.124
620	4,786.6	3,555.6	52.343	1420	13,589.1	10,769.2	61.298
640	4,974.9	3,704.0	52.641	1440	13,834.5	10,974.8	61.469
660	5,165.2	3,854.6	52.934	1460	14,080.8	11,181.4	61.639
680	5,357.6	4,007.2	53.225	1480	14,328.0	11,388.9	61.800
700	5,552.0	4,161.9	53.503	1500	14,576.0	11,597.2	61.974
720	5,748.4	4,318.6	53.780	1520	14,824.9	11,806.4	62.138
740	5,946.8	4,477.3	54.051	1540	15,074.7	12,016.5	62.302
760	6,147.0	4,637.9	54.319	1560	15,325.3	12,227.3	62.464
780	6,349.1	4,800.1	54.582	1580	15,576.7	12,439.0	62.624
800	6,552.9	4,964.2	54.839	1600	15,829.0	12,651.6	62.783
820	6,758.3	5,129.9	55.093	1620	16,081.9	12,864.8	62.939
840	6,965.7	5,297.6	55.343	1640	16,335.7	13,078.9	63.095
860	7,174.7	5,466.9	55.589	1660	16,590.2	13,293.7	63.250
880	7,385.3	5,637.7	55.831	1680	16,845.5	13,509.2	63.403
900	7,597.6	5,810.3	56.070	1700	17,101.4	13,725.4	63.555
920	7,811.4	5,984.4	56.305	1720	17,358.1	13,942.4	63.704
940	8,026.8	6,160.1	56.536	1740	17,615.5	14,160.1	63.853
960	8,243.8	6,337.4	56.765	1760	17,873.5	14,378.4	64.001
980	8,462.2	6,516.1	56.990	1780	18,132.2	14,597.4	64.147
1000	8,682.1	6,696.2	57.212	1800	18,391.5	14,816.9	64.292
1020	8,903.4	6,877.8	57.432	1820	18,651.5	15,037.2	64.435
1040	9,126.2	7,060.9	57.647	1840	18,912.2	15,258.2	64.578
1060	9,350.3	7,245.3	57.861	1860	19,173.4	15,479.7	64.719

T R	$\bar{h}$ Btu/lbmol	$\bar{u}$ Btu/lbmol	$\bar{s}°$ Btu/(lbmol · R)	T R	$\bar{h}$ Btu/lbmol	$\bar{u}$ Btu/lbmol	$\bar{s}°$ Btu/(lbmol · R)
1900	19,698	15,925	64.999	3500	41,965	35,015	73.462
1940	20,224	16,372	65.272	3540	42,543	35,513	73.627
1980	20,753	16,821	65.543	3580	43,121	36,012	73.789
2020	21,284	17,273	65.809	3620	43,701	36,512	73.951
2060	21,818	17,727	66.069	3660	44,280	37,012	74.110
2100	22,353	18,182	66.327	3700	44,861	37,513	74.267
2140	22,890	18,640	66.581	3740	45,442	38,014	74.423
2180	23,429	19,101	66.830	3780	46,023	38,517	74.578
2220	23,970	19,561	67.076	3820	46,605	39,019	74.732
2260	24,512	20,024	67.319	3860	47,188	39,522	74.884
2300	25,056	20,489	67.557	3900	47,771	40,026	75.033
2340	25,602	20,955	67.792	3940	48,355	40,531	75.182
2380	26,150	21,423	68.025	3980	48,939	41,035	75.330
2420	26,699	21,893	68.253	4020	49,524	41,541	75.477
2460	27,249	22,364	68.479	4060	50,109	42,047	75.622
2500	27,801	22,837	68.702	4100	50,695	42,553	75.765
2540	28,355	23,310	68.921	4140	51,282	43,060	75.907
2580	28,910	23,786	69.138	4180	51,868	43,568	76.048
2620	29,465	24,262	69.352	4220	52,456	44,075	76.188
2660	30,023	24,740	69.563	4260	53,044	44,584	76.327
2700	30,581	25,220	69.771	4300	53,632	45,093	76.464
2740	31,141	25,701	69.977	4340	54,221	45,602	76.601
2780	31,702	26,181	70.181	4380	54,810	46,112	76.736
2820	32,264	26,664	70.382	4420	55,400	46,622	76.870
2860	32,827	27,148	70.580	4460	55,990	47,133	77.003
2900	33,392	27,633	70.776	4500	56,581	47,645	77.135
2940	33,957	28,118	70.970	4540	57,172	48,156	77.266
2980	34,523	28,605	71.160	4580	57,764	48,668	77.395
3020	35,090	29,093	71.350	4620	58,356	49,181	77.581
3060	35,659	29,582	71.537	4660	58,948	49,694	77.652
3100	36,228	30,072	71.722	4700	59,541	50,208	77.779
3140	36,798	30,562	71.904	4740	60,134	50,721	77.905
3180	37,369	31,054	72.085	4780	60,728	51,236	78.029
3220	37,941	31,546	72.264	4820	61,322	51,750	78.153
3260	38,513	32,039	72.441	4860	61,916	52,265	78.276
3300	39,087	32,533	72.616	4900	62,511	52,781	78.398
3340	39,661	33,028	72.788	5000	64,000	54,071	78.698
3380	40,236	33,524	72.960	5100	65,491	55,363	78.994
3420	40,812	34,020	73.129	5200	66,984	56,658	79.284
3460	41,388	34,517	73.297	5300	68,471	57,954	79.569

T R	$\bar{h}$ Btu/lbmol	$\bar{u}$ Btu/lbmol	$\bar{s}°$ Btu/(lbmol · R)	T R	$\bar{h}$ Btu/lbmol	$\bar{u}$ Btu/lbmol	$\bar{s}°$ Btu/(lbmol · R)
300	2,081.9	1,486.1	43.223	1080	7,571.1	5,426.4	52.203
320	2,220.9	1,585.4	43.672	1100	7,716.8	5,532.3	52.337
340	2,359.9	1,684.7	44.093	1120	7,862.9	5,638.7	52.468
360	2,498.8	1,783.9	44.490	1140	8,009.2	5,745.4	52.598
380	2,637.9	1,883.3	44.866	1160	8,156.1	5,851.5	52.726
400	2,776.9	1,982.6	45.223	1180	8,303.3	5,960.0	52.852
420	2,916.0	2,081.9	45.563	1200	8,450.8	6,067.8	52.976
440	3,055.0	2,181.2	45.886	1220	8,598.8	6,176.0	53.098
460	3,194.0	2,280.5	46.194	1240	8,747.2	6,284.7	53.218
480	3,333.0	2,379.8	46.491	1260	8,896.0	6,393.8	53.337
500	3,472.1	2,479.2	46.775	1280	9,045.0	6,503.1	53.455
520	3,611.2	2,578.6	47.048	1300	9,194.6	6,613.0	53.571
537	3,725.1	2,663.1	47.272	1320	9,344.6	6,723.2	53.685
540	3,750.3	2,677.9	47.310	1340	9,494.8	6,833.7	53.799
560	3,889.5	2,777.4	47.563	1360	9,645.5	6,944.7	53.910
580	4,028.7	2,876.9	47.807	1380	9,796.6	7,056.1	54.021
600	4,168.0	2,976.5	48.044	1400	9,948.1	7,167.9	54.129
620	4,307.4	3,076.2	48.272	1420	10,100.0	7,280.1	54.237
640	4,446.9	3,175.9	48.494	1440	10,252.2	7,392.6	54.344
660	4,586.6	3,275.8	48.709	1460	10,404.8	7,505.4	54.448
680	4,726.2	3,375.8	48.917	1480	10,557.8	7,618.7	54.522
700	4,886.0	3,475.9	49.120	1500	10,711.1	7,732.3	54.665
720	5,006.1	3,576.3	49.317	1520	10,864.9	7,846.4	54.757
740	5,146.4	3,676.9	49.509	1540	11,019.0	7,960.8	54.858
760	5,286.8	3,777.5	49.697	1560	11,173.4	8,075.4	54.958
780	5,427.4	3,878.4	49.880	1580	11,328.2	8,190.5	55.056
800	5,568.2	3,979.5	50.058	1600	11,483.4	8,306.0	55.154
820	5,709.4	4,081.0	50.232	1620	11,638.9	8,421.8	55.251
840	5,850.7	4,182.6	50.402	1640	11,794.7	8,537.9	55.347
860	5,992.3	4,284.5	50.569	1660	11,950.9	8,654.4	55.411
880	6,134.2	4,386.6	50.732	1680	12,107.5	8,771.2	55.535
900	6,276.4	4,489.1	50.892	1700	12,264.3	8,888.3	55.628
920	6,419.0	4,592.0	51.048	1720	12,421.4	9,005.7	55.720
940	6,561.7	4,695.0	51.202	1740	12,579.0	9,123.6	55.811
960	6,704.9	4,798.5	51.353	1760	12,736.7	9,241.6	55.900
980	6,848.4	4,902.3	51.501	1780	12,894.9	9,360.0	55.990
1000	6,992.2	5,006.3	51.646	1800	13,053.2	9,478.6	56.078
1020	7,136.4	5,110.8	51.788	1820	13,212.0	9,597.7	56.166
1040	7,281.0	5,215.7	51.929	1840	13,371.0	9,717.0	56.253
1060	7,425.9	5,320.9	52.067	1860	13,530.2	9,836.5	56.339

T R	$\bar{h}$ Btu/lbmol	$\bar{u}$ Btu/lbmol	$\bar{s}°$ Btu/(lbmol · R)	T R	$\bar{h}$ Btu/lbmol	$\bar{u}$ Btu/lbmol	$\bar{s}°$ Btu/(lbmol · R)
1900	13,850	10.077	56.509	3500	27,262	20,311	61.612
1940	14,170	10,318	56.677	3540	27,608	20,576	61.710
1980	14,492	10,560	56.841	3580	27,954	20,844	61.807
2020	14,815	10,803	57.007	3620	28,300	21,111	61.903
2060	15,139	11,048	57.161	3660	28,647	21,378	61.998
2100	15,463	11,293	57.317	3700	28,994	21,646	62.093
2140	15,789	11,539	57.470	3740	29,341	21,914	62.186
2180	16,116	11,787	57.621	3780	29,688	22,182	62.279
2220	16,443	12,035	57.770	3820	30,036	22,450	62.370
2260	16,722	12,284	57.917	3860	30,384	22,719	62.461
2300	17,101	12,534	58.062	3900	30,733	22,988	62.511
2340	17,431	12,784	58.204	3940	31,082	23,257	62.640
2380	17,762	13,035	58.344	3980	31,431	23,527	62.728
2420	18,093	13,287	58.482	4020	31,780	23,797	62.816
2460	18,426	13,541	58.619	4060	32,129	24,067	62.902
2500	18,759	13,794	58.754	4100	32,479	24,337	62.988
2540	19,093	14,048	58.885	4140	32,829	24,608	63.072
2580	19,427	14,303	59.016	4180	33,179	24,878	63.156
2620	19,762	14,559	59.145	4220	33,530	25,149	63.240
2660	20,098	14,815	59.272	4260	33,880	25,421	63.323
2700	20,434	15,072	59.398	4300	34,231	25,692	63.405
2740	20,771	15,330	59.521	4340	34,582	25,934	63.486
2780	21,108	15,588	59.644	4380	34,934	26,235	63.567
2820	21,446	15,846	59.765	4420	35,285	26,508	63.647
2860	21,785	16,105	59.884	4460	35,637	26,780	63.726
2900	22,124	16,365	60.002	4500	35,989	27,052	63.805
2940	22,463	16,225	60.118	4540	36,341	27,325	63.883
2980	22,803	16,885	60.232	4580	36,693	27,598	63.960
3020	23,144	17,146	60.346	4620	37,046	27,871	64.036
3060	23,485	17,408	60.458	4660	37,398	28,144	64.113
3100	23,826	17,670	60.569	4700	37,751	28,417	64.188
3140	24,168	17,932	60.679	4740	38,104	28,691	64.263
3180	24,510	18,195	60.787	4780	38,457	28,965	64.337
3220	24,853	18,458	60.894	4820	38,811	29,239	64.411
3260	25,196	18,722	61.000	4860	39,164	29,513	64.484
3300	25,539	18,986	61.105	4900	39,518	29,787	64.556
3340	25,883	19,250	61.209	5000	40,403	30,473	64.735
3380	26,227	19,515	61.311	5100	41,289	31,161	64.910
3420	26,572	19,780	61.412	5200	42,176	31,849	65.082
3460	26,917	20,045	61.513	5300	43,063	32,538	65.252

T R	$\bar{h}$ Btu/lbmol	$\bar{u}$ Btu/lbmol	$\bar{s}°$ Btu/(lbmol · R)	T R	$\bar{h}$ Btu/lbmol	$\bar{u}$ Btu/lbmol	$\bar{s}°$ Btu/(lbmol · R)
300	2,063.5	1,467.7	27.337	1400	9,673.8	6,893.6	37.883
320	2,189.4	1,553.9	27.742	1500	10,381.5	7,402.7	38.372
340	2,317.2	1,642.0	28.130	1600	11,092.5	7,915.1	38.830
360	2,446.8	1,731.9	28.501	1700	11,807.4	8,431.4	39.264
380	2,577.8	1,823.2	28.856	1800	12,526.8	8,952.2	39.675
400	2,710.2	1,915.8	29.195	1900	13,250.9	9,477.8	40.067
420	2,843.7	2,009.6	29.520	2000	13,980.1	10,008.4	40.441
440	2,978.1	2,104.3	29.833	2100	14,714.5	10,544.2	40.799
460	3,113.5	2,200.0	30.133	2200	15,454.4	11,085.5	41.143
480	3,249.4	2,296.2	20.424	2300	16,199.8	11,632.3	41.475
500	3,386.1	2,393.2	30.703	2400	16,950.6	12,184.5	41.794
520	3,523.2	2,490.6	30.972	2500	17,707.3	12,742.6	42.104
537	3,640.3	2,573.9	31.194	2600	18,469.7	13,306.4	42.403
540	3,660.9	2,588.5	31.232	2700	19,237.8	13,876.0	42.692
560	3,798.8	2,686.7	31.482	2800	20,011.8	14,451.4	42.973
580	3,937.1	2,785.3	31.724	2900	20,791.5	15,032.5	43.247
600	4,075.6	2,884.1	32.959	3000	21,576.9	15,619.3	43.514
620	4,214.3	2,983.1	32.187	3100	22,367.7	16,211.5	43.773
640	4,353.1	3,082.1	32.407	3200	23,164.1	16,809.3	44.026
660	4,492.1	3,181.4	32.621	3300	23,965.5	17,412.1	44.273
680	4,631.1	3,280.7	32.829	3400	24,771.9	18,019.9	44.513
700	4,770.2	3,380.1	33.031	3500	25,582.9	18,632.4	44.748
720	4,909.5	3,479.6	33.226	3600	26,398.5	19,249.4	44.978
740	5,048.8	3,579.2	33.417	3700	27,218.5	19,870.8	45.203
760	5,188.1	3,678.8	33.603	3800	28,042.8	20,496.5	45.423
780	5,327.6	3,778.6	33.784	3900	28,871.1	21,126.2	45.638
800	5,467.1	3,878.4	33.961	4000	29,703.5	21,760.0	45.849
820	5,606.7	3,978.3	34.134	4100	30,539.8	22,397.7	46.056
840	5,746.3	4,078.2	34.302	4200	31,379.8	23,039.2	46.257
860	5,885.9	4,178.0	34.466	4300	32,223.5	23,684.3	46.456
880	6,025.6	4,278.0	34.627	4400	33,070.9	24,333.1	46.651
900	6,165.3	4,378.0	34.784	4500	33,921.6	24,985.2	46.842
920	6,305.1	4,478.1	34.938	4600	34,775.7	25,640.7	47.030
940	6,444.9	4,578.1	35.087	4700	35,633.0	26,299.4	47.215
960	6,584.7	4,678.3	35.235	4800	36,493.4	26,961.2	47.396
980	6,724.6	4,778.4	35.379	4900	35,356.9	27,626.1	47.574
1000	6,864.5	4,878.6	35.520	5000	38,223.3	28,294.0	47.749
1100	7,564.6	5,380.1	36.188	5100	39,092.8	28,964.9	47.921
1200	8,265.8	5,882.8	36.798	5200	39,965.1	29,638.6	48.090
1300	8,968.7	6,387.1	37.360	5300	40,840.2	30,315.1	48.257

TABLE A-23E
Ideal-gas properties of water vapor, H_2O

T R	$\bar{h}$ Btu/lbmol	$\bar{u}$ Btu/lbmol	$\bar{s}^°$ Btu/(lbmol · R)	T R	$\bar{h}$ Btu/lbmol	$\bar{u}$ Btu/lbmol	$\bar{s}^°$ Btu/(lbmol · R)
300	2,367.6	1,771.8	40.439	1080	8,768.2	6,623.5	50.854
320	2,526.8	1,891.3	40.952	1100	8,942.0	6,757.5	51.013
340	2,686.0	2,010.8	41.435	1120	9,116.4	6,892.2	51.171
360	2,845.1	2,130.2	41.889	1140	9,291.4	7,027.5	51.325
380	3,004.4	2,249.8	42.320	1160	9,467.1	7,163.5	51.478
400	3,163.8	2,369.4	42.728	1180	9,643.4	7,300.1	51.360
420	3,323.2	2,489.1	43.117	1200	9,820.4	7,437.4	51.777
440	3,482.7	2,608.9	43.487	1220	9,998.0	7,575.2	51.925
460	3,642.3	2,728.8	43.841	1240	10,176.1	7,713.6	52.070
480	3,802.0	2,848.8	44.182	1260	10,354.9	7,852.7	52.212
500	3,962.0	2,969.1	44.508	1280	10,534.4	7,992.5	52.354
520	4,122.0	3,089.4	44.821	1300	10,714.5	8,132.9	52.494
537	4,258.0	3,191.9	45.079	1320	10,895.3	8,274.0	52.631
540	4,282.4	3,210.0	45.124	1340	11,076.6	8,415.5	52.768
560	4,442.8	3,330.7	45.415	1360	11,258.7	8,557.9	52.903
580	4,603.7	3,451.9	45.696	1380	11,441.4	8,700.9	53.037
600	4,764.7	3,573.2	45.970	1400	11,624.8	8,844.6	53.168
620	4,926.1	3,694.9	46.235	1420	11,808.8	8,988.9	53.299
640	5,087.8	3,816.8	46.492	1440	11,993.4	9,133.8	53.428
660	5,250.0	3,939.3	46.741	1460	12,178.8	9,279.4	53.556
680	5,412.5	4,062.1	46.984	1480	12,364.8	9,425.7	53.682
700	5,575.4	4,185.3	47.219	1500	12,551.4	9,572.7	53.808
720	5,738.8	4,309.0	47.450	1520	12,738.8	9,720.3	53.932
740	5,902.6	4,433.1	47.673	1540	12,926.8	9,868.6	54.055
760	6,066.9	4,557.6	47.893	1560	13,115.6	10,017.6	54.117
780	6,231.7	4,682.7	48.106	1580	13,305.0	10,167.3	54.298
800	6,396.9	4,808.2	48.316	1600	13,494.4	10,317.6	54.418
820	6,562.6	4,934.2	48.520	1620	13,685.7	10,468.6	54.535
840	6,728.9	5,060.8	48.721	1640	13,877.0	10,620.2	54.653
860	6,895.6	5,187.8	48.916	1660	14,069.2	10,772.7	54.770
880	7,062.9	5,315.3	49.109	1680	14,261.9	10,925.6	54.886
900	7,230.9	5,443.6	49.298	1700	14,455.4	11,079.4	54.999
920	7,399.4	5,572.4	49.483	1720	14,649.5	11,233.8	55.113
940	7,568.4	5,701.7	49.665	1740	14,844.3	11,388.9	55.226
960	7,738.0	5,831.6	49.843	1760	15,039.8	11,544.7	55.339
980	7,908.2	5,962.0	50.019	1780	15,236.1	11,701.2	55.449
1000	8,078.2	6,093.0	50.191	1800	15,433.0	11,858.4	55.559
1020	8,250.4	6,224.8	50.360	1820	15,630.6	12,016.3	55.668
1040	8,422.4	6,357.1	50.528	1840	15,828.7	12,174.7	55.777
1060	8,595.0	6,490.0	50.693	1860	16,027.6	12,333.9	55.884

T R	$\bar{h}$ Btu/lbmol	$\bar{u}$ Btu/lbmol	$\bar{s}°$ Btu/(lbmol · R)	T R	$\bar{h}$ Btu/lbmol	$\bar{u}$ Btu/lbmol	$\bar{s}°$ Btu/(lbmol · R)
1900	16,428	12,654	56.097	3500	34,324	27,373	62.876
1940	16,830	12,977	56.307	3540	34,809	27,779	63.015
1980	17,235	13,303	56.514	3580	35,296	28,187	63.153
2020	17,643	13,632	56.719	3620	35,785	28,596	63.288
2060	18,054	13,963	56.920	3660	36,274	29,006	63.423
2100	18,467	14,297	57.119	3700	36,765	29,418	63.557
2140	18,883	14,633	57.315	3740	37,258	29,831	63.690
2180	19,301	14,972	57.509	3780	37,752	30,245	63.821
2220	19,722	15,313	57.701	3820	38,247	30,661	63.952
2260	20,145	15,657	57.889	3860	38,743	31,077	64.082
2300	20,571	16,003	58.077	3900	39,240	31,495	64.210
2340	20,999	16,352	58.261	3940	39,739	31,915	64.338
2380	21,429	16,703	58.445	3980	40,239	32,335	64.465
2420	21,862	17,057	58.625	4020	40,740	32,757	64.591
2460	22,298	17,413	58.803	4060	41,242	33,179	64.715
2500	22,735	17,771	58.980	4100	41,745	33,603	64.839
2540	23,175	18,131	59.155	4140	42,250	34,028	64.962
2580	23,618	18,494	59.328	4180	42,755	34,454	65.084
2620	24,062	18,859	59.500	4220	43,267	34,881	65.204
2660	24,508	19,226	59.669	4260	43,769	35,310	65.325
2700	24,957	19,595	59.837	4300	44,278	35,739	65.444
2740	25,408	19,967	60.003	4340	44,788	36,169	65.563
2780	25,861	20,340	60.167	4380	45,298	36,600	65.680
2820	26,316	20,715	60.330	4420	45,810	37,032	65.797
2860	26,773	21,093	60.490	4460	46,322	37,465	65.913
2900	27,231	21,472	60.650	4500	46,836	37,900	66.028
2940	27,692	21,853	60.809	4540	47,350	38,334	66.142
2980	28,154	22,237	60.965	4580	47,866	38,770	66.255
3020	28,619	22,621	61.120	4620	48,382	39,207	66.368
3060	29,085	23,085	61.274	4660	48,899	39,645	66.480
3100	29,553	23,397	61.426	4700	49,417	40,083	66.591
3140	30,023	23,787	61.577	4740	49,936	40,523	66.701
3180	30,494	24,179	61.727	4780	50,455	40,963	66.811
3220	30,967	24,572	61.874	4820	50,976	41,404	66.920
3260	31,442	24,968	62.022	4860	51,497	41,856	67.028
3300	31,918	25,365	62.167	4900	52,019	42,288	67.135
3340	32,396	25,763	62.312	5000	53,327	43,398	67.401
3380	32,876	26,164	62.454	5100	54,640	44,512	67.662
3420	33,357	26,565	62.597	5200	55,957	45,631	67.918
3460	33,839	26,968	62.738	5300	57,279	46,754	68.172

TABLE A-26E
Enthalpy of formation, Gibbs function of formation, and absolute entropy at 77°F, 1 atm

Substance	Formula	$\bar{h}_f^\circ$ Btu/lbmol	$\bar{g}_f^\circ$ Btu/lbmol	$\bar{s}^\circ$ Btu/(lbmol · R)
Carbon	C(s)	0	0	1.36
Hydrogen	$H_2(g)$	0	0	31.21
Nitrogen	$N_2(g)$	0	0	45.77
Oxygen	$O_2(g)$	0	0	49.00
Carbon monoxide	CO(g)	−47,540	−59,010	47.21
Carbon dioxide	$CO_2(g)$	−169,300	−169,680	51.07
Water vapor	$H_2O(g)$	−104,040	−98,350	45.11
Water	$H_2O(l)$	−122,970	−102,040	16.71
Hydrogen peroxide	$H_2O_2(g)$	−58,640	−45,430	55.60
Ammonia	$NH_3(g)$	−19,750	−7,140	45.97
Methane	$CH_4(g)$	−32,210	−21,860	44.49
Acetylene	$C_2H_2(g)$	+97,540	+87,990	48.00
Ethylene	$C_2H_4(g)$	+22,490	+29,306	52.54
Ethane	$C_2H_6(g)$	−36,420	−14,150	54.85
Propylene	$C_3H_6(g)$	+8,790	+26,980	63.80
Propane	$C_3H_8(g)$	−44,680	−10,105	64.51
n-Butane	$C_4H_{10}(g)$	−54,270	−6,760	74.11
n-Octane	$C_8H_{18}(g)$	−89,680	+7,110	111.55
n-Octane	$C_8H_{18}(l)$	−107,530	+2,840	86.23
n-Dodecane	$C_{12}H_{26}(g)$	−125,190	+21,570	148.86
Benzene	$C_6H_6(g)$	+35,680	+55,780	64.34
Methyl alcohol	$CH_3OH(g)$	−86,540	−69,700	57.29
Methyl alcohol	$CH_3OH(l)$	−102,670	−71,570	30.30
Ethyl alcohol	$C_2H_5OH(g)$	−101,230	−72,520	67.54
Ethyl alcohol	$C_2H_5OH(l)$	−119,470	−75,240	38.40
Oxygen	O(g)	+107,210	+99,710	38.47
Hydrogen	H(g)	+93,780	+87,460	27.39
Nitrogen	N(g)	+203,340	+195,970	36.61
Hydroxyl	OH(g)	+16,790	+14,750	43.92

Source: From the JANAF, *Thermochemical Tables*, Dow Chemical Co., 1971; *Selected Values of Chemical Thermodynamic Properties*, NBS Technical Note 270-3, 1968; and *API Research Project 44*, Carnegie Press, 1953.

Enthalpy of combustion and enthalpy of vaporization at 77°F, 1 atm
(Water appears as a liquid in the products of combustion)

Substance	Formula	$\Delta \bar{h}_c^{\circ} = -$**HHV** **Btu/lbmol**	$\bar{h}_{fg}$ **Btu/lbmol**
Hydrogen	$H_2(g)$	−122,970	
Carbon	$C(s)$	−169,290	
Carbon monoxide	$CO(g)$	−121,750	
Methane	$CH_4(g)$	−383,040	
Acetylene	$C_2H_2(g)$	−559,120	
Ethylene	$C_2H_4(g)$	−607,010	
Ethane	$C_2H_6(g)$	−671,080	
Propylene	$C_3H_6(g)$	−885,580	
Propane	$C_3H_8(g)$	−955,070	6,480
n-Butane	$C_4H_{10}(g)$	−1,237,800	9,060
n-Pentane	$C_5H_{12}(g)$	−1,521,300	11,360
n-Hexane	$C_6H_{14}(g)$	−1,804,600	13,563
n-Heptane	$C_7H_{16}(g)$	−2,088,000	15,713
n-Octane	$C_8H_{18}(g)$	−2,371,400	17,835
Benzene	$C_6H_6(g)$	−1,420,300	14,552
Toluene	$C_7H_8(g)$	−1,698,400	17,176
Methyl alcohol	$CH_3OH(g)$	−328,700	16,092
Ethyl alcohol	$C_2H_5OH(g)$	−606,280	18,216

Source: Kenneth Wark, Thermodynamics, 3d ed., McGraw-Hill, New York, 1977, p. 879, table A-23.

TABLE A-29E
Constants that appear in the Beattie-Bridgeman and the Benedict-Webb-Rubin equations of state

(a) The Beattie-Bridgeman equation of state is

$$P = \frac{R_u T}{\bar{v}^2}\left(1 - \frac{c}{\bar{v}T^3}\right)(\bar{v} + B) - \frac{A}{\bar{v}^2} \quad \text{where} \quad A = A_0\left(1 - \frac{a}{\bar{v}}\right) \quad \text{and} \quad B = B_0\left(1 - \frac{b}{\bar{v}}\right)$$

When P is in psia, $\bar{v}$ is in $ft^3/lbmol$, T is in R, and $R_u = 10.73 \text{ psia} \cdot ft^3/(lbmol \cdot R)$, the five constants in the Beattie-Bridgeman equation are as follows:

Gas	A_0	a	B_0	b	c
Air	4,905.096	0.3093	0.7386	−0.017 64	4.054×10^6
Argon, Ar	4,865.515	0.3729	0.6297	0.0	5.596×10^6
Carbon dioxide, CO_2	18,872.857	1.142	1.678	1.159	6.166×10^7
Helium, He	81.424	0.9587	0.2243	0.0	3.737×10^3
Hydrogen, H_2	744.510	−0.081 05	0.3357	−0.6982	4.708×10^4
Nitrogen, N_2	5,068.324	0.4192	0.8083	−0.1107	3.924×10^6
Oxygen, O_2	5,620.956	0.4104	0.7407	0.067 41	4.484×10^6

Source: Computed from Table A-29a by using the proper conversion factors.

(b) The Benedict-Webb-Rubin equation of state is

$$P = \frac{R_u T}{\bar{v}} + \left(B_0 R_u T - A_0 - \frac{C_0}{T^2}\right)\frac{1}{\bar{v}^2} + \frac{bR_u T - a}{\bar{v}^3} + \frac{a\alpha}{\bar{v}^6} + \frac{c}{\bar{v}^3 T^2}\left(1 + \frac{\gamma}{\bar{v}^2}\right)e^{-\gamma/\bar{v}^2}$$

When P is in atm, $\bar{v}$ is in $ft^3/lbmol$, T is in R, and $R_u = 0.730 \text{ atm} \cdot ft^3/(lbmol \cdot R)$, the eight constants in the Benedict-Webb-Rubin equation are as follows:

Gas	a	A_0	b	B_0	c	C_0	α	γ
n-Butane, C_4H_{10}	7747	2590	10.27	1.993	4.219×10^9	8.263×10^8	4.531	8.732
Carbon dioxide, CO_2	563.1	703.0	1.852	0.7998	1.989×10^8	1.153×10^8	0.3486	1.384
Carbon monoxide, CO	150.7	344.5	0.676	0.8740	1.387×10^7	7.124×10^6	0.5556	1.541
Methane, CH_4	203.1	476.4	0.868	0.6827	3.393×10^7	1.878×10^7	0.5120	1.541
Nitrogen, N_2	103.2	270.6	0.598	0.6529	9.713×10^6	6.706×10^6	0.5235	1.361

Source: Kenneth Wark, *Thermodynamics*, 4th ed., McGraw-Hill, New York, 1983, p. 864, table A-21. Originally published in H. W. Cooper and J. C. Goldfrank, *Hydrocarbon Processing*, vol. 46, no. 12, p. 141, 1967.

Psychrometric chart at 1-atm total pressure. (From the American Society of Heating, Refrigerating and Air-Conditioning Engineers; used with permission.)

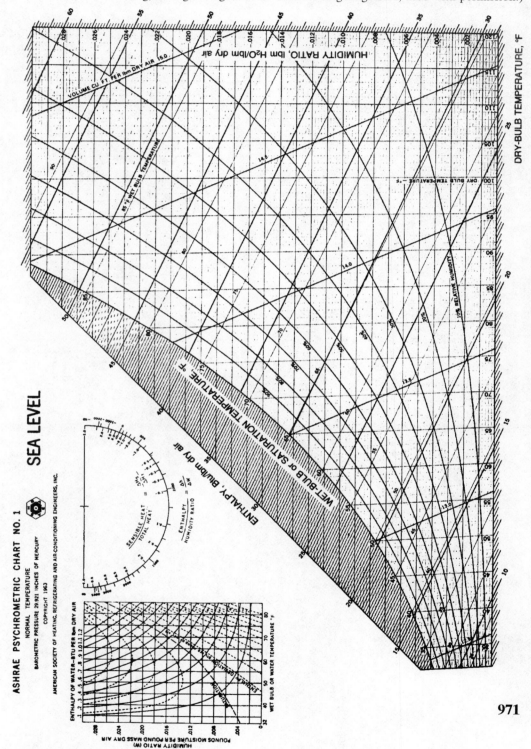

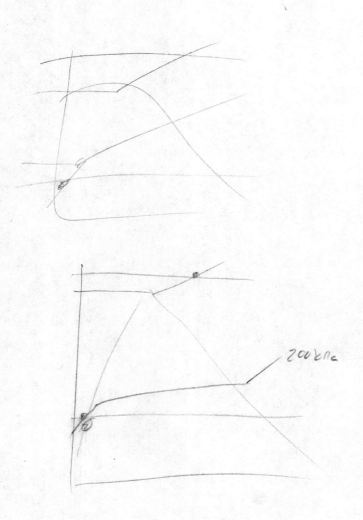

200 kPa